NEUROIMMUNITY

NEUROIMMUNITY

A New Science That Will
Revolutionize How We Keep Our
Brains Healthy and Young

MICHAL SCHWARTZ

with Anat London

With a foreword by Olle Lindvall

Yale UNIVERSITY PRESS/NEW HAVEN & LONDON

Published on the foundation established in memory of William Chauncey Williams of the Class of 1822, Yale Medical School, and of William Cook Williams of the Class of 1850, Yale Medical School.

Yale University Press books may be purchased in quantity for educational, business, or promotional use. For information, please e-mail sales.press@yale.edu (U.S. office) or sales@yaleup.co.uk (U.K. office).

Designed by Mary Valencia.
Set in Perpetua and ITC Stone Sans types by Integrated Publishing Solutions.
Printed in the United States of America.

Library of Congress Control Number: 2015932302
ISBN 978-0-300-20347-9 (cloth : alk. paper)

A catalogue record for this book is available from the British Library.

The information and suggestions contained in this book are not intended to replace the services of your physician or caregiver. Because each person and each medical situation is unique, you should consult your own physician to get answers to your personal questions, to evaluate any symptoms you may have, or to receive suggestions for appropriate medications.
The author has attempted to make this book as accurate and up-to-date as possible, but it may nevertheless contain errors, omissions, or material that is out of date at the time you read it. Neither the author nor the publisher has any legal responsibility or liability for errors, omissions, out-of-date material, or the reader's application of the medical information or advice contained in this book.

This paper meets the requirements of ANSI/NISO Z39.48-1992
(Permanence of Paper).

10 9 8 7 6 5 4 3 2 1

I would like to dedicate this book to my beloved family: my husband, Professor Michael Eisenbach, in gratitude for his true friendship and partnership, and endless support, help, and understanding, with whom I have shared the joy and the difficulties of this scientific endeavor; and my beloved children, Orit, Osnat, Eyal, and Tomer, who experienced with me the "highs" that accompany the joys of discovery, as well as the days, nights, months, and years of hard work, and frequent disappointment. I also dedicate this work to my dear brother, Nathan Hevrony, who has been my soulmate and friend. Last but not least, to my children-in-law, who have learned to accept their unconventional mother-in-law, and to my loving grandchildren.

CONTENTS

FOREWORD

In 2003, four colleagues and I published in the *Proceedings of the National Academy of Sciences of the United States of America* the article "Inflammation is detrimental for neurogenesis in adult brain." We showed that microglia/macrophages activated after an insult to the brain (status epilepticus) compromise the survival of new hippocampal neurons soon after they have been born. Our findings were completely in line with the consensus among scientists at that time—namely, that inflammation in the central nervous system (CNS) is harmful after injury or disease. Immune cells were regarded as inactive under normal conditions and dangerous to the brain under pathological conditions. Consequently, therapeutic interventions aimed at suppressing immune responses. However, as Michal Schwartz so clearly describes in this book, and as is now accepted by the scientific community, the action of the immune system is much more complex. A paradigm shift has taken place in recent years, and it is now generally believed that immune cells are important to maintaining normal brain function, and that they can be beneficial for repair processes. A dysfunctional immune system may cause impairments in cognition and mood and contribute to inadequate regeneration following injury and progression of neurodegenerative disease.

Michal Schwartz has pioneered and continues to have the leading role in this change of our view on the interaction between the immune system and the brain and spinal cord. Introducing completely novel ideas in science, then taking them from the initial stage of skepticism and rejection toward final acceptance, is a major challenge. Success in this endeavor requires unique scientific creativity as well as courage and persistence. This book gives us the opportunity to follow Michal Schwartz on her exciting scientific journey: fifteen years of groundbreaking experimental studies, always from the perspective of how her findings might be used in human disease. We see several examples of frontline research that proved both thought-provoking and hypothesis-generating, inspiring work in many other laboratories.

Because the world's population is getting older, discoveries of how the immune system seems to repair the aging brain and counteract decline of cognitive functions such as learning and memory acquire ever-increasing impact. Aging of the immune system contributes to cognitive dysfunction. By restoring the biochemical equilibrium, the immune system also helps the brain to cope with stress, which otherwise may lead to long-lasting mental dysfunction.

When my colleagues and I published our 2003 paper on the subject, inflammation was regarded as harmful and was thought to require suppression. As part of the paradigm shift described in this book, we now have evidence that different steps in the recovery process following injury involve distinct subclasses of immune cells, which switch on and off in sequence and on schedule to successfully accomplish their tasks. Based on this new knowledge, Schwartz and her collaborators have developed a protocol in which a subset of immune cells, macrophages, are delivered at the appropriate time. The macrophages secrete molecules that can support tissue repair and bring the inflammatory process under control. This so-called macrophage therapy has already reached Phase 1 clinical trials.

A recent seminal finding, which is described in several chapters, is that the choroid plexus is an active immunological organ, sensor, and gateway, secreting molecules and allowing selected immune cells to enter the central nervous system for repair. For many years, the choroid plexus was known for producing cerebrospinal fluid (CSF) and acting as a filtration system, removing metabolic waste, foreign substances, and excess neurotransmitters from the CSF. In Alzheimer's disease, the choroid plexus loses some of its properties, leading to a failure to recruit appropriate immune cells to the CNS. The choroid plexus gate is shut in Alzheimer's patients. The immune system produces more suppressive cells, which block the immune response. One way to overcome the immune suppression could be through vaccination to safely boost the protective immune response.

What makes Schwartz's book particularly interesting is the perspective of the individual researcher as she builds her case regarding brain-immune system interaction: the ideas and visions, the ups and downs, and the resistance of and final acceptance by the scientific community. Her account also is a valuable contribution to the history of science, illustrating the thrill of the scientific process when one can posit a novel hypothesis, develop the most elegant and conclusive experimental design, and ultimately win acceptance from other researchers for the paradigm shift.

Much work remains, though, before the role of the immune system for normal brain function, as well as in disease and injury, is completely understood. Better knowledge of basic mechanisms is necessary so that the full potential of the exciting findings in animal models can be implemented in the clinic. The extent to which this field has developed, especially in its clinical applications, calls to mind Winston Churchill's stirring words following an early Allied victory in World War II: "Now this is not the end. It is not even the beginning of the end. But it is, perhaps, the end of the beginning." I am convinced that the scientific journey described in this book will continue, for Michal Schwartz and her collaborators, and for many other researchers around the world.

Olle Lindvall
Professor of Neurology
Lund University
Lund, Sweden

PREFACE

In 2008 a multidisciplinary conference dealing with the concept of memory from the perspectives of philosophers, scientists, psychologists, physicists, and historians was held in Mishkenot Sha'ananim, one of Jerusalem's most active and sophisticated cultural institutions, overlooking the Old City walls. When I presented my new understanding of how who we are reflects what we remember, and how what we remember is affected by our immune system, the audience was surprised by the concept. However, upon further reflection, many in the audience commented that my theory made tremendous sense, and I was asked why this understanding had not become common knowledge. I explained that what now seemed obvious was actually the outcome of years of research going against consensus and the common dogma. I immediately realized that I was very used to writing for my scientific peers but had never attempted to describe my theory for the general reader. It was then that the idea of writing this book was born. My research has since reached the point where the pieces that constitute my theory have taken shape as a story of interest to individuals wishing to understand more about the brain and how they could improve their cognitive performance and their well-being.

The English poet Percy Bysshe Shelley wrote, "The more we study, the more we discover our ignorance." Although years of intense research have provided some answers, they have raised just as many new questions. I believe I can now provide evidence of the mechanisms that underlie my initial discoveries, as well as some new insights into the mystery of what is needed to maintain the optimal function of our brains over the years.

My training in both immunology and neuroscience has facilitated my exploration of how the immune system and the mind—two pivotal systems of the body that previously have been considered as completely separate—actually act in harmony to preserve brain health. Briefly, I propose that the immune system—the body's network of cells and organs that defends us against disease—is the key player that keeps the brain in good condition. This theory offers a way to harness the immune system to maintain optimal brain health throughout life, proposing new hope for treating various neurological conditions, and even for rolling back the aged brain to a more youthful state.

When I finished my doctoral studies as an immunologist under the supervision of Professors Michael Sela and Edna Mozes at the Weizmann Institute of Science, it was believed that although the immune system is essential for repairing all the body's other tissues, the brain was excluded. This commonly held belief was based on the function of the blood-brain barrier, which allows optimal maintenance of the brain's stable milieu for its delicate and accurate everyday performance. To me, the idea that an essential organ such as the brain, which is not transplantable, has "given up" such invaluable assistance from the immune system seemed almost paradoxical. This was the point at which I decided to join a research group, headed by Professor Bernard Agranoff at the University of Michigan, Ann Arbor, that at that time (1978) was leading the field of central nervous system repair. It was in this laboratory that I was first confronted with the inability of the central nervous system to heal following damage.

When I returned to the Weizmann Institute, armed with my enhanced understanding of the brain and the immune system, I felt that the time was ripe to test my idea that the immune system is indeed needed for brain and spinal cord repair. I reasoned that the limited ability of the brain to benefit from the immune system is not a purposeful evolutionary choice but rather an outcome of an evolutionary trade-off in which the accessibility to immune system assistance is limited in pathology for the sake of optimal brain performance in health. Still, it took me several years to reach the scientific stage that enabled me to pursue this line of research and to make it my life's work.

Since 1998, my team at the Weizmann Institute has been able to provide experimental evidence suggesting that the immune system and the brain "speak" to each other. We pioneered the idea that the immune system plays an indispensable role in keeping your brain healthy and functioning well throughout your life. When this system fails, dormant brain diseases can emerge, and cognitive decline begins to be manifested.

Challenging conventional dogma wasn't easy. As is often the case in science, thinking outside the box earned me many critics. As I started on this journey, many of my colleagues warned me that I was making a huge mistake, entering a scientific cul-de-sac that could waste years and ruin my reputation. At times, I thought my colleagues were right; it seemed as if I was getting nowhere. As the data accumulated and more pieces were added to the puzzle, however, I felt more confidence in my theory, and I continued to move forward, seeking greater in-depth understanding. I was surprised to discover that the brain's dependency on the immune system is more robust and broader than I could ever have imagined.

Now, long after my initial findings were published in 1998, it is clear that the leap of scientific imagination that set me on this journey—supported by generations of outstanding students who joined me, believed in my concept, and helped to build and strengthen it—has brought us closer to our destination. My job is now to extend our research in the hope of finding cures for many diseases of the brain. The immune system that was once perceived as the villain is now increasingly accepted as a key player in the repair process, one that should be modulated rather than thwarted.

When I decided to write this summary of my scientific endeavor, I approached Anat London, a former graduate student with whom I had had many hours of scientific discussion, and with whom I enjoyed thinking and writing.

"It was during my undergraduate studies that I first heard of Michal Schwartz," Anat recalls. "I attended a lecture by Professor Benjamin Sredni, a former chief scientist of Israel's Ministry of Health, and the director of the Cancer, AIDS, and Immunology Research (CAIR) Institute of Bar-Ilan University, Israel. Sredni talked about cancer and the immune response; he discussed the way the immune system determines whether we get sick and how we heal. He introduced Michal Schwartz: 'A pioneering researcher at the Weizmann Institute of Science.' He talked about the way she had gone bravely against the common wisdom by showing that the immune system, then thought not to interact with the brain, actually defends the brain from the development of various pathologies and is essential for the daily maintenance of the brain. It was so fascinating; I knew then that I must meet this researcher."

When Anat started her master's degree at the Weizmann Institute of Science, she came to meet me, and we decided she would start to work at my laboratory for a short internship. "This short internship eventually became seven-year journey," Anat recalls, "during which I have had the privilege of working with Schwartz and taking part in her scientific endeavor.

"I have been inspired and motivated by the enthusiasm Schwartz had for her research, by her endless passion for knowledge and continuous striving for scientific achievements that could potentially benefit mankind."

Anat was excited to hear of my long-held desire to share my discoveries beyond the scientific community. We decided to write this book together and to make my scientific theory approachable for others who wish to know more about how the brain keeps its health, how to improve their mental health, and how to strengthen their brain and mind by using the power of the immune system.

ACKNOWLEDGMENTS

The scientific work covered by this book could not have been done without the contributions of outstanding students who joined me in the course of my research. With some of them I have spent days, evenings, and late nights sharing, arguing, fighting, and letting our imaginations test the limits of the possible. Such long discussions led each generation of students to make a new quantum leap in understanding. Each leap required another five years to prove and to create the additional milestones that make up this book.

My special thanks go to Eti Yoles, Ehud Hauben, Jonathan Kipnis, Yaniv Ziv, Noga Ron Harel, Asya Rolls, Gil Lewitus, Michal Cardon, Ravid Shechter, Anat London, Ana Catarina

Raposo, Inbal Benhar, Kuti Baruch, Gilad Kunis, and Aleksandra Deczkowska, who contributed to the conceptual milestones that assembled the puzzle presented in the book, and to other students without whose dedication and hard work this story would not have materialized.

I also wish to thank my beloved brother Nathan Hevrony, a successful businessman and a wonderful human being, with whom I share many of my thoughts, and who believed in me. Together we decided to make all efforts to translate the know-how for the benefit of people everywhere. Special thanks to my dear husband, Professor Michael Eisenbach, who has given me the strength to continue in this bumpy journey and has supported me through all the years of long hours, dedication, and agony. Last but not least, my beloved children, Orit, Osnat, Eyal, and Tomer, who while undoubtedly proud of my achievements, must have felt for many years that they were competing for my affections with science and with my research and students.

Most of the scientific journey described in this book took place at the Weizmann Institute of Science, one of the world's leading multidisciplinary research institutions, which provides an outstanding infrastructure and scientically enriched environment. This book took shape with the help of many talented people, including Genia Brodsky, who helped with the graphic design of the figures, and Ann Downer-Hazell, who edited the book. I wish to thank Professor Avraham A. Levy and members of my team, who shared with me their valuable thoughts and comments.

INTRODUCTION

S cience, technology, and medicine have made almost un-
imaginable strides in the past century, eradicating disease
and premature death. In 1913, the average American man
could expect to live to fifty years of age, the average American
woman five years longer. By 2013, American men could expect to live an average of seventy-seven years, women eighty-
two, a gain of almost thirty years, with comparable increases
in the United Kingdom and Europe. The United States ranks
thirty-fifth on the longevity list; Monaco, Japan, and Iceland
have some of the highest life expectancies: A woman born in
Monaco can expect to celebrate her ninetieth birthday. We are
living longer, healthier lives than at any time in human history.

Therein lies a paradox. As we live longer, we fear the ravages of old age. For the seventy-eight million baby boomers born after World War II, as for many people around the world entering their senior years, aging is not always a welcome prospect. For many people of this generation, the mere idea of getting old is repugnant, and they are willing to pay a great deal of money for the privilege of staying young. We don't just fear wrinkles and a decreased sex drive; we fear infirmity and dementia, as well as loss of identity, independence, and the ability to make decisions for ourselves. In addition, depression and age-related brain diseases such as Alzheimer's and Parkinson's threaten our mental health.

Your brain, the part of the body that determines who you are, is a network of interconnected biological wires (neurons) that serves as your interface with the world around you. It gathers a flood of sensations, organizes it into a clear message, and dictates the way you respond to this message. It determines how you learn and remember, how you behave, feel, and interact with others. It shapes the essence of your personality. Yet the brain is constantly subjected to challenges, to which it must adapt to ensure optimal function. Otherwise, we are predisposed to a host of conditions that can impair our thinking and mental stability. If they persist, these conditions may lead to depression, age-related dementia, and chronic diseases such as Alzheimer's and Parkinson's.

In the chapters that follow, we will introduce the revolutionary theory that the immune system keeps the brain in good condition. We will present new insights into the mechanism by which the brain keeps itself young despite decades of wear and tear; how the brain maintains its peak performance; and how it restores equilibrium after any activity that tilts its balance toward the positive (listening to music) or the negative (enduring stress). We will show how new therapies might be developed based on these new insights. By sharing them with you, we want to show how these discoveries in brain sciences and immunology have opened up new avenues of understanding that allow anyone to maintain a healthier brain and mind.

In the past, the brain was considered an autonomous organ. It was viewed as self-contained tissue, cut off from the immune system for its defense, maintenance, and repair. This view was based largely on the brain's structure and on interpretation of observations made of diseased brains.

In fact, most of what we know about the way the healthy body functions emerges from observations made of patients suffering from disease, or those who have otherwise lost function. This often leads scientists to misinterpretations, to treating the body's response to harmful conditions as if it were destructive; they sometimes ignore the possibility that the body's response may reflect attempts at repair, and that these attempts may fall short or fail.

Brain pathologies have also been viewed in this way, so immune cells recruited following brain damage were traditionally considered part of the problem; from that developed the generalization that immune cells should not be allowed in the brain under any circumstances, and from that arose the widespread and often erroneous use of steroids to weaken the immune system.

In this book you will explore the discoveries that overturned this misconception. My journey that turned a long-held dogma on its head has been long and personally painful, but the end justifies the efforts. The new discoveries made by my team have influenced research toward finding remedies for some fatal brain diseases. The discoveries have also changed the way some clinicians and scientists handle and treat brain injury, spinal cord injury, and chronic diseases like Alzheimer's disease and amyotrophic lateral sclerosis (known as ALS or Lou Gehrig's disease).

The studies presented in this book have shown that the brain and the immune system are two closely intertwined whole-body systems, not tangential, as researchers had long believed. The "wireless" immune system, though it is not part of the wired system of the brain's neuronal network, is essential for providing an optimal framework for proper brain function and repair.

In 1999 I published the theory of protective autoimmunity, which proposes that the immune system that recognizes the brain is essential for the maintenance of a healthy brain, eyes, and spinal cord. Today this theory is more widely accepted by the scientific community, including some of the scientists who were initially my most vociferous critics. The question is no longer *whether* immune cells keep the brain healthy and play an integral part of its function, but *how* they do so, and how we can make sure that we most effectively utilize neuroimmunity throughout life, during stressful times and especially as we age.

This book will take you through the ups and downs of my research. You will stand beside us at the laboratory bench, seeing how our experiments unfolded. We will share my team's ideas as they took shape, our triumphs and setbacks, and the moments of surprise that led to some of the exciting revelations in the field of brain health and repair. Finally, we will share with you the thoughts and plans for applying this knowledge to make our lives better and healthier. We hope this book will allow you to better understand how your own immune system benefits your nervous system, and how to maintain the immune system in a way that will improve your life and possibly protect you from disease. Although often taken for granted, the brain is one tissue that we will never be able to replace. The heart, kidney, liver, and other tissues are now transplantable. No matter how much technology and knowledge advance, the brain and spinal cord will never be fully replaceable. Therefore we ought to decipher the mechanisms by which our body keeps the brain and the spinal cord functioning, despite years of wear and tear, and learn to exploit such mechanisms for therapies. This book will help teach you such strategies.

In the primer at the back of this book you will find an accessible summary of the scientific terms and concepts that will provide you with the basics needed to follow the rest of the book. If you need an introduction to the basics of neural anatomy and the immune system, start with this primer. If you already know the basics of the brain and the immune system, you can begin with Chapter 1.

Note: This book is a summary of my scientific endeavor that led to the revolutionized view of the relationships between the immune system and the mind. I wrote this book with Anat London, a former graduate student and a current scientific consultant at my laboratory. My personal views and anecdotes are written in the first person singular. The first person plural refers sometimes to my research team, sometimes to Anat London and me as authors, sometimes to humanity as a whole; the context of the material should make the pronoun reference clear.

ABBREVIATIONS

ADHD	attention deficit/hyperactivity disorder
ALS	amyotrophic lateral sclerosis
AMD	age-related macular degeneration
BDNF	brain-derived neurotrophic factor
CNS	central nervous system
GFP	green fluorescent protein
NGF	nerve growth factor
PTSD	posttraumatic stress disorder
RRMS	relapsing-remitting multiple sclerosis
SOD1	superoxide dismutase 1

NEUROIMMUNITY

1

A New Player in the Body-Mind Connection

The Immune System

In the early 1630s, René Descartes was wandering the streets of Amsterdam, calling at various butcher shops. He wasn't shopping for dinner. Rather, he took home the haunches of beef and other cuts of meat to dissect. Descartes, a Frenchman famous for his contributions to philosophy and mathematics, was also a keen student of anatomy. When he could get them, he also dissected human cadavers. On his mind was the question, Where does the soul reside? Does it animate the body? Or is the body more like a machine?

In his *Treatise of Man,* published posthumously in 1664, Descartes argues that human physiology and behavior follow the same principles as the physical universe. He saw the body as a machine similar to "clocks, artificial fountains, and mills," which operated according to physical rules dependent "solely on the disposition of our organs." He set out an explanation of a system of nerve fibers transmitting messages to and from the brain, and was the first person to propose that an involuntary action could be produced by sensory stimulation—a revolutionary idea that would acquire the name "reflex" a century later.

But with respect to complex behaviors, Descartes was still a product of the seventeenth century. His science was deeply influenced by the Catholic Church's doctrine of the soul. From his point of view, we pick up a pencil only after our soul decides to do so and orders the muscles to perform the action. Descartes coined the term "dualism" as a way of describing the mind and body as two different entities, which interact in a small structure in the brain, the pineal gland. Although his concept of a complex system that delivers and receives sensory stimulation and accordingly dictates the body's response was correct, he erroneously described this system as one that is dominated by "animal spirits" that transfer the sensory information—a belief that Gilbert Ryle three hundred years later called "the dogma of the ghost in the machine."[1]

The pendulum of thought about where our consciousness resides has shifted throughout the years: from viewing the "soul" as being located outside the physical body, we have come to find consciousness part of our physical existence residing in the brain, which is separated from the rest of the body but controls it. This question of where the self resides has been a contentious subject for many centuries. In this chapter we will see how philosophers and physicians addressed the brain and mind throughout history and how they viewed the body-mind interaction. We will offer a twist in the adventure story by suggesting a new and unexpected player in the body-mind equation—the immune system. If, we now understand, the brain controls our intellectual, emotional, and physical performance through its network of neurons, the immune system resets the "workplace" in which these neurons are operating, the brain's milieu, in order to provide an optimal framework for proper brain function without interference. In other words, the immune system is not part of the brain's network of neurons but is the body's orchestrating system, which controls the brain. We have proposed that the immune system is the central coordinating mechanism that enables the entire physiology of mind and body to function in harmony.

HISTORICAL PERSPECTIVE: WHERE DOES THE MIND RESIDE?

In the late 1700s, German physician and neuroanatomist Franz Joseph Gall was among the first to propose that each behavior is controlled by specific brain regions. According to his view there was a spot in the brain linked to any behavior you could think of, including generosity, the sense of satire and wit, the ability to keep secrets, and the talents for mathematics, architecture, and music. Gall's view was controversial, as some philosophers and intellectuals were reluctant to accept the idea that each mental function had a specific physiological basis.

The concept of no soul, no spirits in the machine, was too much for them to bear.[2]

The desire to unravel the way our behavior is controlled did not start with Descartes or Gall. The Edwin Smith papyrus (c. 1600 BC) contains the earliest known reference to the brain, describing the symptoms and prognosis of patients with head injuries. However, the Egyptians and some of the Greek philosophers in the centuries to follow did not think much of the brain. This may explain why Egyptian priests carefully preserved the heart and other organs during mummification while disposing of the brain. Philosophers such as Aristotle perceived the heart as the seat of the mind, responsible for our emotions and thoughts. The heart has retained a metaphorical connection to our emotional states, in expressions like "broken heart" or "change of heart." Other philosophers, such as Pythagoras, Hippocrates, and Plato, viewed the brain, rather than the heart, as the source of our intellect, emotions, and reasoning.[3]

The scholastic tradition in which Descartes was educated was largely based on the biomedical texts written by the Greek surgeon Galen, the most influential and appreciated philosopher at the time. Galen served as the in-house physician for school of Roman gladiators, a job that provided the young doctor with extensive experience in treating head trauma. Galen witnessed the supreme role of the brain in controlling both movement and behavior. He himself was among the Greeks who believed that the mind and body were a single entity. By the time Descartes was making his first steps in deciphering the relationship between the mind and the body, these issues of the mind or soul versus the physical body, and what role the brain played in the interaction between the two, were still a matter of contention.[4]

Since Descartes's doctrine, breakthroughs in technique and technology have allowed researchers to understand that the nervous system is responsible for our behavior and to decipher the way it works. In the nineteenth century, Camillo Golgi, an Italian scientist, developed a silver stain that provided the breakthrough needed to visualize neurons in the brain. Later, Santiago Ramón y Cajal, the Spanish anatomist who became the father of contemporary neuroscience, improved Golgi's method and mapped an individual nerve cell. Ramón y Cajal showed that neurons were single, complete cells, though of a complex physical nature. He showed that each neuron contained a cell body, fingerlike dendrites that collect the information and deliver it into the cell body, and a long axon that extends from the cell body and carries information to the edge of the neuron toward the next target cell (figure 1).[5]

Ramón y Cajal's discoveries made him perhaps the greatest neuroscientist of all time. His findings became the basis for all brain research that followed. As a young man, he had wanted to be an artist, and, using Golgi's silver stain, he produced a series of drawings of nerve cells that are striking in their beauty.

In 1923 he wrote an elegant description of his search for the secrets of the inner workings of the brain in which he described the "cells of delicate and elegant forms" as "the mys-

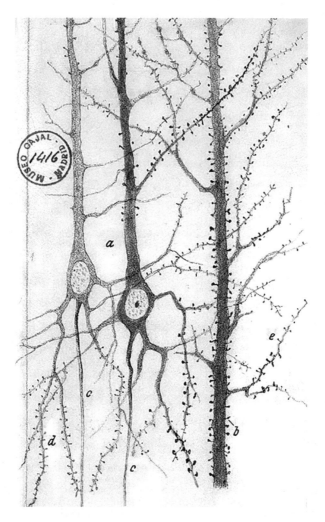

Figure 1. Original drawing by Ramón y Cajal showing brain neurons of a rabbit (1896, black ink and pencil). Fingerlike dendrites are clearly depicted. Ramón y Cajal Legacy. Instituto Cajal (CSIC). Madrid (Spain).

terious butterflies of the soul, whose fluttering wings would some day—who knows?—enlighten the secret of mental life." In the same spirit, if the neurons are the mysterious butterflies of the soul, our findings led me to suggest that the immune cells are the nectar of these butterflies—maintaining their functions and keeping them safe.

Contemporary with Ramón y Cajal was the Austrian neurologist Sigmund Freud, the founding father of psychoanalysis and a highly skilled physician with a deep understanding of the physical structure of the brain. One of Freud's early scientific papers, published in 1882, was a study of the nerve cells of crayfish; without knowing the Spanish scientist, he independently reached the same conclusion as Ramón y Cajal: that the nerve cell is a single unit.

Freud's deep insight into human psychology was to recognize that our current emotional life and personality are intimately connected to our parents, to our past experience, and to our environment. He claimed that he was able to treat patients' problems by encouraging them to freely discuss their thoughts and memories, a practice that led to the establishment of what was then an entirely new branch of medicine. In the current age of drug therapies for every mental condition, classical Freudian psychoanalysis may have lost some of its prestige. Despite his background as a physician and the fact that he was an expert in neuroanatomy, Freud could not explain his observations and consequential hypotheses in physical terms. The id, ego, and superego could not be located in the brain. Two hundred years after Descartes, Freud had still not recognized the body-mind connection. The ghost had still not been exorcised from the machine.[6]

The attribution of specific behaviors to different brain parts was advanced in the mid-nineteenth century by British neurologist J. Hughlings Jackson. While studying epilepsy, he provided evidence that different parts of the brain controlled diverse motor and sensory functions. Later on, Ramón y Cajal and others showed that different mental functions are regulated by certain groups of neurons within the brain that are connected to one another.[7]

With the help of new tools and techniques that were not available to Ramón y Cajal and Freud, contemporary neuroscience has developed a much deeper understanding of the physical processes involved in brain activity, which are responsible for many activities of the mind. The American neuropsychiatrist Eric Kandel, who was awarded the Nobel Prize in physiology and medicine in 2000, has studied the interactions of neurons in the humble central nervous system of a marine snail. Together with many scientists of his generation, he has provided new insights into the cellular and molecular mechanisms that underlie learning and memory formation. Advances in technology allow scientists to listen to the electrical impulse as it travels along an axon and identify the molecules that enable neurons to transmit signals along neural pathways. Particular mental functions have been associated with specific areas of the brain; at the same time, researchers have seen how an injured brain can shift these functions to another area. New techniques have enabled us to trace firing neurons in real time in the brains of living animals and humans in a technique called functional magnetic resonance imaging, or fMRI.[8]

DOES THE IMMUNE SYSTEM LINK BODY AND MIND?

This philosophical debate continued for centuries over the concept of a body-mind dualism and the question as to whether the mind/soul was contained within the body's physical properties. Now that, thanks to the major development in neurosciences, many functions of the mind have been discovered to be located within the brain, and body-mind dualism has taken on new meaning. Researchers are currently conducting experiments to tease apart how the brain is able to perform so many sophisticated and diverse tasks, accurately, precisely, and over the lifespan of the individual. New questions emerge: Is the mind assisted by the body outside the brain to maintain its health for so many decades? Does the body outside the brain affect the brain, or is the brain totally autonomous?

Descartes and his contemporaries, and many others to follow, in debating the body-mind connection, never considered the immune system in this equation. This is not surprising; at that time in the history of science, the immune system had not been identified. Research that would lead to germ theory was still in its infancy, and recognition of the role of the immune system in tissue maintenance was far in the future. As late as the mid-nineteenth century, people believed "miasmas" caused disease. Progress came slowly. In the early eighteenth century, Mary Wortley Montagu, the wife of the British ambassador to Turkey, inoculated her own children after witnessing local doctors in Istanbul inoculating children with pus from mild smallpox cases against the deadly smallpox epidemic. In the 1790s Edward Jenner noted that milkmaids infected with cowpox were largely immune to smallpox, and he used the abscess of their cowpox lesions to vaccinate healthy individuals, who, when they were later exposed to smallpox, did not develop the disease.

Later, Louis Pasteur and Robert Koch demonstrated that microbes can cause diseases. Their insights in turn led to the understanding of cellular immunity, antibody function, and the development of synthetic antibiotic therapy.

For most of the twentieth century, immunologists still continued to view the immune system as one that is solely designed by nature to protect us against invading threatening pathogens. Toward the end of the century additional roles of the immune system have been revealed, including its roles in normal maintenance of tissues, healing noninfectious "sterile" wounds, and coping with internal enemies, such as hazardous molecules produced by our own bodies, or our own cells if they become cancerous. Step by step, more elements of the immune system were identified, and its role in maintaining and defending the body was revealed. Yet the brain was completely excluded from this discussion, largely because of its unique anatomy as a tissue behind barriers.[9]

When I entered the field of neuroscience as a young post-doc after earning my Ph.D. in immunology, I felt that it was highly unlikely that such a precious and sophisticated system as the central nervous system would have given up the maintenance of the immune system.

My students and I were the first to ask, Could the immune system also protect the brain? Could it affect the way we think and feel, or even the way we adapt to the endless changes and challenges that we encounter throughout our life? Could it be the system that repairs the brain, the most crucial organ, from the wear and tear that accumulates throughout life?

Even when the immune system was recognized as the ultimate system of maintenance and repair, it was not thought to be shared among all organs. It was long believed that some parts of our body are not assisted by immune cells so readily. Among these restricted organs, also called immune privileged sites, are the brain, the spinal cord, and the neural retina: the components of the central nervous system. Experiments in the early 1920s demonstrated that tumor cells transplanted into the brain can grow and survive for prolonged periods, compared with tumors placed beneath the skin, which are rapidly attacked by immune cells. These experiments further supported the idea that immune cells, which generally act to reject foreign transplants, are not active in the brain under normal conditions. A generation after these experiments, the brain still was considered a sealed system, surrounded by barriers that screened out most blood-borne cells and molecules.[10]

Because scientists used to believe that immune cells were kept out of the brain, the fields of neuroscience and immunology long remained separate. At its outset, the study of brain and immunity was a field that focused solely on how to prevent immune cells from attacking the brain and other parts of the central nervous system in various pathologies, as in the case of multiple sclerosis, a disease in which immune cells attack the brain. Thus the possible benefits of the immune system in the central nervous system were not even considered. Instead, the immune system was viewed as a major threat to the brain.

However, my team began to ask questions. Had evolution really left the brain walled off and deprived of help from the immune system? This question opened up new avenues in the research of brain-immune interactions.

My students and I were the team that overturned this tenet in 1998, by proposing that the brain had retained the ability to be assisted by the immune system. At that time, it was already accepted that the mind is part of the physical body. We were adding another key component to the body-mind puzzle, showing that the immune system is the way our body keeps the brain and mind in good condition.

COULD IMMUNE CELLS SHAPE THE MIND?

Since 1998 my group's research has suggested a much closer interdependence between the physical body and the inner workings of the mind. Our work has revealed that the brain has close relations to the immune system. We showed that the immune system plays such a central role in the function and maintenance of the brain that it is nothing less than the "nectar" that feeds Ramón y Cajal's "mysterious butterflies."

The dawn of the twenty-first century has brought a series of experiments that have changed neuroscience. Our own studies revealed that immune cells control formation of the brain's stem cells, shape cognitive performance such as learning and memory, and affect our mood and our ability to cope with stress. In all these ways, the immune system helps in shaping the mind. Although such immune cells are not regular citizens of the brain, and they are not an integral part of the wired system, we now know that they are located in special borders of the brain. When the brain calls for help, immune cells assist it remotely via molecules that are delivered to the brain, or by their controlled entry to help repair the diseased brain.[11]

Our studies offer new insights into the body-mind relationship. In the past the question was whether the mind is located within the body, in the brain, as part of our physical existence, and if so, how it affects the body. We were now asking the question the other way around: Can the mind function properly over decades without the assistance of the body outside the brain? As you will see, we are suggesting that the fitness of our body dictates the fitness of our mind, affecting the way we think and reason, the way we learn and express emotions. It does so via the immune system. In the chapters to follow we will show you how (figure 2).

These new discoveries have paved the way for a new uni-

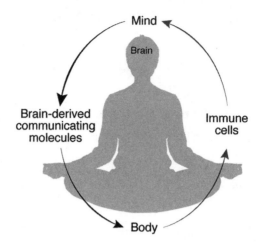

Figure 2. Immune cells support the mind.

fied theory that views the immune system as a key player in the physiology of the brain and mind, whose breakdown may cause various central nervous system pathologies, including to the brain, spinal cord, and eyes. This unified theory, unveiled here systematically for the first time, prompted us to look at brain and spinal cord pathologies as well as mental conditions in a new way. It inspired us to investigate neurological pathologies such as depression, age-related cognitive decline, Alzheimer's and Parkinson's diseases, head and spinal cord injuries, and many other central nervous system disorders while focusing on the immune system, trying to shed light on its natural defense functions. Could we harness our immune cells as a potential therapy to treat disease, and prevent immune system deterioration as an anti-aging strategy? We set out to try.

CHAPTER

2

Cognition and the Aging Brain
The Immune Cells of Wisdom

O ur enormous capabilities to learn and remember, to
create stories, to compose symphonies, and to design
buildings and spaceships all depend on the ability of the brain
to adapt to the constantly changing environment. This ability
is known as brain plasticity. In recent decades, it has become
clear that this plasticity is dependent on the brain's capacity to
make new connections between neurons, to strengthen existing connections among these cells, and to make new neurons
throughout our life, a process termed neurogenesis. However,
the ability of the adult brain to form new neurons was a foreign concept at the time of Ramón y Cajal, the godfather of
modern neuroscience, who claimed, "In the adult brain, nerv-

ous pathways are fixed and immutable; everything may die, nothing may be regenerated."[1]

In the early 1960s this dogma was challenged for the first time by the American biologist Joseph Altman. Though he observed the formation of new neurons in the adult brain, and published his results in the prestigious journal *Science,* his discoveries were ignored for decades. Finally, Elizabeth Gould, Fred H. Gage, and other scientists overturned this tenet, demonstrating that new neurons are formed throughout life in the hippocampus, the part of the brain responsible for learning and memory. Further studies revealed that formation of new neurons in the adult brain is essential for learning and coping with stress, plays a role as the brain ages, and affects brain diseases such as Alzheimer's disease and depression.[2]

In searching for factors that affect the formation of new neurons, researchers were intrigued to find that balanced, not compulsive, physical activity increases the number of new neurons formed in the brain.[3] Puzzled by this power of the body on the mind, my team and I wondered how the brain translates physical activity into instructions to make new neurons. We asked which factors are created in the body during physical activity that could be so influential on the mind. In 2006, my graduate students at that time, Jonathan Kipnis (now a professor at the University of Virginia) and Yaniv Ziv (now a senior scientist at the Weizmann Institute of Science), and I speculated that the immune system might play the role of middleman between the body and the brain during physical activity.

Our speculation regarding the possibility that the immune system links physical activity and neurogenesis was reinforced by the knowledge at that time that, on one hand, the formation of new neurons is associated with the mechanism of action of antidepressants and, on the other hand, doctors and therapists, as a way of fighting depression, often recommended physical activity. We were therefore intrigued by the idea that perhaps we would discover the mysterious players that assemble these various pieces of the puzzle. We loved the idea that it could be the immune system that helps translating physical activity into the formation of new neurons, which consequently lift our mood. We were left with the question, How?

In testing this hypothesis, we asked whether a normally functioning immune system supports the creation of new neurons. We compared two groups of mice: one with a compromised immune system, such that they resembled "bubble boys" (patients with a congenital immune deficit who are kept in isolation to avoid coming into contact with people who might infect them with bacteria and viruses), and a group of mice with normally functioning immune systems. We quantified the numbers of newly formed neurons in the brains of these two groups of mice. Immune-deficient mice formed significantly lower numbers of new neurons, indicating the profound way immune cells influence brain neurogenesis (figure 3). When we restored the immune systems of the defective mice by transplanted immune cells, in a procedure much like that performed in leukemia patients who need to undergo

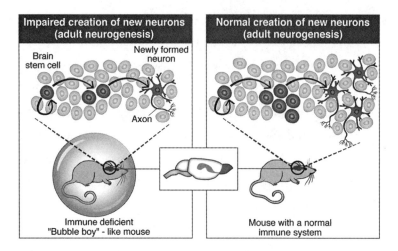

Figure 3. Immune cells support the creation of new neurons in the brain. "Bubble boy"–like mice, born with immune deficiency similar to babies born with congenital immune deficiency, have lower numbers of newly formed neurons in areas associated with learning and memory, such as the hippocampus.

bone-marrow transplantation, we were able to show increased numbers of newly formed nerve cells, similar to the numbers we found in mice with normal immune systems. Our results indicated, for the first time, that appropriate manipulation of the immune system outside the brain could benefit processes occurring within the brain responsible for cognition.

Excited by the strong influence of the immune system on the mind, we wondered whether animals that have defective immune systems could enjoy the benefit of physical activity on their brain as normal mice do. We allowed the mice to live in cages that provided them with an enriched environment—cages that included a variety of objects, toys, a set of tunnels, and running wheels in which the mice were motivated to perform enhanced physical activities. Subsequently, we monitored the levels of neurogenesis in their brains. The immune-defective mice showed almost no benefit from the physical activity, whereas the mice with the healthy immune systems showed, as expected, elevated levels of new neurons. These results suggested that the physical activity conveys messages to the brain that are communicated via the immune system. When the immune system malfunctions, this communication channel can't work effectively.[4] No wonder that running, yoga, and other types of athletic workout have become preferred activities during our leisure time. Now we know that they can make us smarter by strengthening the immune system that our mind needs in order to stay sharp (figure 4).

Even more striking, our studies further revealed, using several animal models, that no matter how healthy the brain is, its ability to learn and retain memories, and its attention capacity, are influenced by the health of the immune system. In evaluating cognitive capacity, we tested the animals in a task that required each mouse to learn and memorize its position in space. In one of these tasks, called the Morris Water Maze, the animal is placed into a round pool, which contains a platform under the surface that is not seen by the mouse when it is placed in the water. Mice don't like water, and once the mouse reaches the platform, it stands on it and looks around to learn its dimensions, and how to escape from the water. After a few days, the animal is placed back in the same pool, and it is tested for the time it spends around the location of the

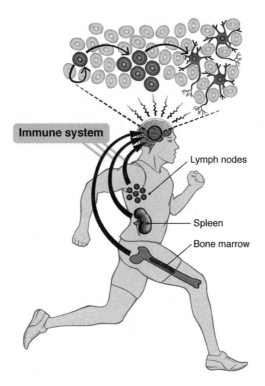

Figure 4. Exercise boosts the immune system, which in turn supports the formation of new neurons in the brain (neurogenesis).

platform, the time it takes to reach the platform, and the pathway it chooses to get there (whether it swims directly toward the platform or discovers it randomly). Again we compared "bubble boy"–like mice born with defective immune systems with a group of mice with normally functioning immune systems. Our researchers performed these trials in a "blinded" fashion, which means that they did not know whether the mice in each test group had functioning immune systems or not. At some point they phoned me to say that one of the groups of mice was extremely stupid. It turned out to be the immune-deficient group! The mice whose immune systems were un-compromised easily learned and remembered the location of the platform, but the immune-deficient mice were unable to complete the cognitive task. They seemed to swim randomly; they spent less time in the area in which the platform was originally placed and more time just searching for the plat-form (figure 5). Once we restored the immune functions of such mice, by immune cell transplantation, they were able to perform the task significantly better. This was the first time that anyone had demonstrated that cognitive performance de-pends on cells *outside* the brain, and that the brain is not as autonomous as was previously thought. Of course, these cells of the immune system are not part of the wired system of the brain, and are not directly involved in the electrical activity of the brain required for cognitive performance. Nonetheless, it seems that without the support of such immune cells the brain can't function properly. These results stunned us; why should our intellectual performance depend on cells other than the cells within the brain?

We also observed that animals born with immune defi-ciency displayed an attention deficit and not only learning and memory deficits. The most surprising results were that, although the animals were born with compromised immune systems, their behavioral impairment (attention deficit) man-ifested only when they reached puberty. These results were

Normal immune system Immune deficient

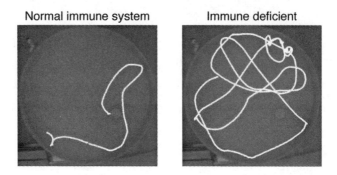

Figure 5. Immune cells support cognitive functions. In a learning and memory test (Morris Water Maze), mice are placed in a water pool and are trained to find a hidden platform that enables them to escape the water. Immune deficient mice, right, cannot remember the location of the platform and continue to swim around randomly; mice with working immune systems, left, remember the location of the platform and swim directly to it.

even more striking, because they immediately reminded us of what we knew about neurodevelopmental diseases, like schizophrenia, a disease that is often linked to the aberrant development of the fetus during pregnancy but emerges around puberty. The psychotic symptoms that we saw in the mice that were born with immune deficit were reminiscent of those found in schizophrenic patients. It was exciting to discover that an immune deficit causes attention deficits! Even more exciting was when we realized that immune cell transplantation could reverse these deficits. If in the past we all thought that "bubble boys" were defective only with respect to their ability to fight against bacteria and viruses, now we understood that they might also suffer from cognitive and behavioral impairment. The behavioral impairment is reversible: Just as transplantation of bone marrow in these children builds up their immune systems and makes them resilient to infections, we now understand that it may also restore cognitive ability, alleviate attention deficits, and build up resilience to stress, as will be discussed in Chapter 3. Thus, although we knew that immune cells don't determine our intelligence (IQ) or emotional intelligence (EI), we were excited to learn that we may make the most of our brain potential by strengthening our immune system or preventing its decay. Such understanding raised an urgent need to find how the immune system does it.[5]

The powerful influence of immune cells on cognitive ability drove us to think that these cells must be of a unique type to perform such a sophisticated task. We found that these cells recognize brain components. In this study we used mice that were genetically engineered so that almost all their lymphocytes (T cells) could recognize only brain proteins. Such self-recognizing cells are often called autoimmune cells.

In 1960, a Nobel Prize was awarded to Sir Frank Macfarlane Burnet, best known for proposing that during the development of our own immune systems, immune cells that recognize the body's self compounds—autoimmune cells—are deleted to ensure that no such self-recognizing immune cells exist in the body, as these cells might later in life cause autoimmune diseases. Since then scientists have considered such immune cells to be hazardous to the body as well as to the brain, and they were only associated with autoimmune diseases. The thought that some autoimmune cells that are found in all individuals, without causing disease, might actually exist in the body for its own benefit has not been accepted by the scientific community. The results we describe in this book evidently support the position that not all autoimmune cells are "enemies"; some of them are the "miraculous substance" of our brain.

I still remember the day, in July 1997, when I was walking in the gardens of the Weizmann Institute and I had this thought, came back with excitement about the idea, and had a hard time persuading the graduate students to do the experiment. It was Eti Yoles, at that time research associate in my laboratory with a strong background in physiology but not in immunology, who picked up the gauntlet and conducted the experiment that was reproduced and published in the 1999 paper in which we introduced the concept.[6]

When we evaluated how new neurons were formed in mice that had been genetically engineered to have almost exclusively autoimmune cells, which recognize the body's compounds, they displayed increased formation of new neurons compared with normal mice, which had fewer autoimmune cells. The mice that were born enriched with cells that recognize brain compounds—the brain's specific autoimmune cells—also performed better in the learning and memory test, the Morris Water Maze, than did mice with normal immune system repertoires. Thus our studies showed that autoimmune cells that recognize components of the brain not only are harmless but are even essential if we hope to maintain the brain in good condition, generate new nerve cells, and preserve cognitive functions such as learning and memory. We now call these brain-recognizing immune cells "the immune cells of wisdom," and will thus refer to them throughout the book.[7]

As we became convinced that the immune system is essential for day-to-day brain performance, we suspected, as you will see below, that the aging brain is influenced by our aging immune system. As we age, are we losing "the immune cells of wisdom"?

IS AGING EPIDEMIC?

Diseases implicated in past epidemics, like smallpox, polio, and other major killers, have been all but eradicated in the developed world thanks to immunization and more effective treatment. Women with access to modern medical care seldom die in childbirth. Advances in treating diseases mean that some diagnoses that meant early death one hundred years ago are now chronic diseases that can be managed. Personalized genetic medicine promises treatment tailored to each individual patient's genome.

Safety has advanced, too. Workplace safety laws and robots employed for dangerous jobs make workplaces safer. Far fewer people die in car accidents since the advent of seat belts and air bags. Fire deaths have plummeted thanks to widespread use of sprinkler systems, smoke detectors, and fire-retardant fabrics.

All these factors help us live longer, but as we live longer we fear more than ever the consequences of old age. The seventy-eight million baby boomers born after World War II are obsessively searching for ways to hold back the clock. To many people of this generation, the mere idea of getting old is repugnant, and they are willing to pay a great deal of money for the privilege of staying young. By 2015 the anti-aging industry is expected to generate more than $291 billion in revenue worldwide. This includes hormone replacement therapy, cosmetic surgery, skin care products and treatments, medications to treat erectile dysfunction, customized multivitamin and herbal preparations, and many other products costly to our wallets and occasionally also to our health.[8]

WHAT IS CURRENTLY IN THE PIPELINE TO FIGHT AGING OF THE BODY?

Before we address aging and the mind, let us briefly review what is currently available to fight aging in general. When scientists speak of aging, they mean the accumulation of damage in cells, tissues, and organs culminating in cell death and tissue deterioration. Wrinkles, erectile dysfunction, arthritis, dementia, and all other symptoms we associate with aging are just the consequence of the wear and tear on our cells and tissues.[9]

The aging process manifests itself in every system of our body. We get wrinkles after years of exposure to ultraviolet wavelengths of sunlight, different pollutants (especially tobacco smoke), and even the oxygen in the air. All these factors, together with the common contemporary condition of high blood sugar levels, activate chemical reactions that create free radicals (unstable molecules that attack components of our cells), and degrade connective tissues in our skin, further aging us. Our hair turns gray as our hair follicles deteriorate and don't produce as much melanin as they did when we were young. We lose our sex drive partly because our ovaries or testes secrete lower levels of sex hormones. We get painful, stiff joints when our cartilage wears away, making our bones rub against each other. We become forgetful and even demented as deposits of protein fragments and fibers build up between and within our nerve cells, disrupting nerve communication, and leading to cell death in brain areas essential for learning and memory. Modern aging researchers focus on finding ways to repair the existing damage, to delay or arrest further deterioration, and to extend cell life.

So what can we do to halt and even reverse the wear and tear of the mind? Before I share with you my theory on rejuvenation of the brain and mind, let's first review current research worldwide on reversing erosion of the body outside the central nervous system.

Several research directions hold promise. One of them entails restoring hormones to youthful levels. Hormone replacement therapy, or HRT, has over the years had inconsistent support in the scientific community, as its safety has received conflicting assessments.[10]

Another approach, calorie restriction, shows promise both in laboratory animals and in some human populations. Human trials have shown that eating fewer calories while maintaining proper nutrition can reduce the risk for thickening of the arteries. It can also relieve aging-associated dysfunction of the heart, allow maintenance of a healthy weight, and stave off diabetes. One natural laboratory in which these effects are evident is Okinawa, a tiny group of islands off the coast of Japan. Okinawans live longer and have the highest concentration of centenarians in the world. However, those who leave Okinawa and adopt the typical American diet of processed food have shorter lifespans, similar to those of the general U.S. population, indicating that their longevity is largely dependent on their lifestyle and related to what they eat and the number of calories they consume. The ongoing Okinawa Centenarian Study, established in 1975, reveals a more complex explanation to this amazing longevity phenomenon by identifying "human longevity genes" in centenarians and other elderly people living in Okinawa. Thus the study supports the notion that both genetic and environmental factors matter, and that the interactions among these factors ultimately leads to healthy longevity.[11]

An additional research direction is gene therapy, which involves silencing age-related genes and inserting chosen genes that can improve health and extend lifespan. A recent animal study showed that mice that were treated by a gene responsible for delaying cell death lived longer than mice that didn't receive the gene. This gene codes for a protein called telomerase, which elongates the endings of the chromosomes that get shorter as we age. Eventually, these shortened chromosomes lead to cell aging and death. A Nobel Prize was awarded for the discovery of this protein in 2009 to Elizabeth H. Blackburn, Carol W. Greider, and Jack W. Szostak.[12]

Although this approach is appealing, major safety issues remain to be resolved. For example, the risk of cancer increases with telomerase therapy, as cancerous cells tend to exploit telomerase to promote uncontrollable cell division, leading to tumors. An additional safety issue involves the reaction of the patient's immune system to the carrier of the gene, which is an engineered virus used to deliver the gene of interest. The immune reaction evoked by the virus was probably the cause of the unfortunate result of a gene therapy clinical trial conducted in 1999: the death of Jesse Gelsinger, who suffered from an X chromosome–linked genetic disease of the liver. Gelsinger, eighteen, volunteered in a clinical trial that attempted to treat a genetic disorder from which he suffered with a virus that delivers a functional gene instead of the defective one. Four days after Gelsinger received the gene therapy, he was dead; gone with him was the entire field of gene therapy, or so it seemed at the time. In 1999 the number of gene therapy trials in the United States was at its peak, but after Gelsinger's death that number had plummeted to thirty-four by 2001. Since then, new viruses that induce minimal response have been identified and cataloged. Slowly but consistently, the field of gene therapy is rebuilding its credentials as a legitimate therapeutic approach.[13]

Stem cell therapy to regenerate tissue and replace lost cells

29

is another research field drawing increased attention among researchers in the field of aging. Stem cells have the potential to form most of the cells of the adult body and thus may assist in healing many age-related disorders. As opposed to adult stem cells, each of which is committed to becoming a certain cell type and can usually regenerate the tissue from which it derived, embryonic stem cells are more versatile and can form almost all cell types. Human embryonic stem cells are harvested from embryos, which are destroyed in the process, posing ethical concerns. In addition, the use of such cells bears a risk of transplant rejection, necessitating the heavy use of drugs to suppress immune-rejection activities. An additional type of stem cell that holds promise and might serve as an alternative to embryonic stem cells is the induced pluripotent stem cell (iPSC). These cells were first identified in 2006 by Shinya Yamanaka, who was awarded the 2012 Nobel Prize for his discovery (together with John B. Gurdon, who discovered in the 1960s that specialization of cells is reversible). iPSCs are adult cells that are genetically reprogrammed to acquire a stem cell–like state, and therefore can potentially give rise to various types of cells. As in the field of gene therapy, safety issues—such as stem cells' ability to grow in the wrong place or to form the wrong tissue, to induce cancer, or to trigger a robust response by the immune system—should be considered before their use in humans.[14]

We now know that if any of these promising regimens affects the immune system, it should also affect the mind, since the mind is so much influenced by the immune system. The question is whether and how the mind can benefit from a more direct manipulation of the immune system.

BUT WHAT CAN BE DONE TO KEEP OUR MINDS YOUNG?

Even if we manage to extend our life expectancy, will it guarantee that we can maintain our brains in good shape? According to an old saying, "Anyone who stops learning is old, whether twenty or eighty. Anyone who keeps learning stays young. The greatest thing you can do is keep your mind young." Unfortunately, like the rest of our body, the brain also changes as we age. These changes include the shrinkage of brain cells and the inability to rearrange connections between cells, needed for any brain's activity, especially learning and memory. In addition, brain tissue can be damaged by oxidants, molecules that react to create cell-threatening substances, such as free radicals. As we age, abnormal deposits appear between and within nerve cells, waste products collect in the brain tissue, and DNA damage accumulates in our cells that can no longer adequately repair it.

These changes affect every aspect of the lives of the elderly. Delivery and processing of information by nerve cells becomes less efficient, which leads to slower reaction times that can in turn make driving hazardous; poor eyesight and hearing that may interfere with work, family, social life, and hobbies. Integration of feedback about posture and gait from the eyes, inner ear, and muscles is impaired, leading to loss of balance, falls, and injuries. Age-related brain changes can lead to mental confusion that makes the elderly easy prey for financial scams and identity thieves, disrupts their ability to plan and make decisions, and interferes with their adjustment to the constantly changing world.

Moreover, as the brain ages, neurodegenerative diseases such as Alzheimer's disease (in about five million Americans age sixty-five and older), Parkinson's disease (in about one million Americans age fifty and older), and the cognitive decline associated with dementia can develop.[15]

Scientists have only a limited understanding of why aging decreases cognitive ability and increases mental dysfunction. For decades the common belief was that the brain ages when brain neurons are lost. It was accepted by neuroscientists and gerontologists alike that neurons die while no new neurons are added throughout our lifespans. As we have seen, this accepted wisdom has been overturned; we now know that new neurons arise constantly in specific areas of the adult brain. In addition, behavioral symptoms of aging often precede the death of neurons. Thus we should be able to devise therapies that intervene long before neurons are lost in order to halt irreversible memory loss and attention deficits. As long as the cells are there, we can prevent their deterioration. The question is how. To achieve this goal, we need to understand whether aging of the immune system is involved in aging of the mind, and if so, whether we can rejuvenate the immune system to rejuvenate the mind.

THE AGING IMMUNE SYSTEM

As described at the beginning of this chapter, we need a well-functioning immune system to make the best of our brain potential—to perform cognitive tasks and routine brain functions throughout our lifetime. Does the immune system age with the rest of the body? Is it possible that the immune system shows signs of aging *before* the brain? Could this be the leading factor in brain aging?

Unfortunately, the aging process of the body does not spare the immune system. As we age, the immune system deteriorates in a process known as immune senescence. Researchers have identified a wide range of age-related changes in the immune system, including imbalance between effector immune cells and suppressor cells (see the primer in the back of the book). In aging, the balance tilts in favor of cells that suppress the immune system, leaving the individual with reduced numbers of immune cells and lower levels of antibodies. In the aged body, the thymus, an organ accountable for immune cell development and education, atrophies. This results in decreased output of functional immune cells. In addition, the hematopoietic components of the bone marrow, responsible for continuous production of immune cells, are replaced by fat tissue. Age-related changes in the architecture of immune organs are also evident. Together, these changes combine to cause lowered immune responses in the elderly.[16]

The decline of the immune response with age, at a time when the brain's maintenance services are most needed, prompted us to consider the possibility that the aging immune system might contribute to age-related "brain fog" and other signs of cognitive decline. In other words, we did not think that immune aging *causes* brain aging but rather that it determines when symptoms of aging are manifested in the brain and the speed with which brain aging progresses. The brain undergoes its own wear and tear, and we found that the immune system is in charge of providing repair to the brain and restoration of balance whenever it is lost. The brain steps onto the slippery slope of deterioration—cognitive aging—only when it accumulates damage through this erosion and the immune system not only can no longer meet the brain's needs but even loses its ability to help repair that damage.

We first asked what would happen if a young individual experienced a reduction in immunity, similar to that usually experienced by an elderly individual. Would his or her cognitive abilities be affected? Several lines of evidence indicate that weakening of the immune system impairs cognitive functions. AIDS patients, whose immune system is under attack by the human immunodeficiency virus (HIV), experience symptoms like reduced attention and concentration, delay in speech and verbal fluency, and even HIV-related dementia. Cancer patients undergoing chemotherapy that weakens the immune system experience similar declines in various brain functions.

We artificially manipulated the immune response in eight-week-old mice that had reached sexual maturity to simulate immune deficiency that occurs in aging. We observed a decline in cognitive functions. Together with Noga Ron-Harel, a Ph.D. student at that time (now a postdoctoral fellow at Harvard), we tested the mice in the same water maze described above and found that the young mice with healthy brains but with artificially aged immune systems were capable of learning the location of the platform but did not remember its location when placed back in the pool after a while, indicating a memory deficit.

We also used another cognitive test for spatial memory, called a novel location recognition task, to test these young mice in which we artificially aged the immune system. In this task the mice are placed in a square open field and familiarized with two small objects located at two corners of the field. After a while, one of the objects is placed in a different corner. Mice are curious creatures, and they ordinarily prefer to explore objects in novel locations rather than in familiar ones. Of course, to show this novelty preference a mouse must be able to remember the previous location and recognize it. Mice with normal cognitive abilities will remember the previous location of the object that has been moved. They will spend less time around the object that was left in its original location and show preference for exploring the object that was moved to a new location. We measure this preference by noting the time the mouse spends close to the object, sniffing and touching it with the tip of its nose or with its paws. Compared with mice with a normal immune system, young mice with the artificially aged immune systems showed a profound loss in their preference for exploring an object in its novel location, again indicating a memory shortage (figure 6). We concluded that the age of the immune system, not the individual's chronological age, was a key factor in mental sharpness.[17]

But can we dial back the immune system and reverse age-related cognitive decline? The first thing that came to mind was simply to replace the old immune system with a younger model. As immune cells are constantly produced in the bone marrow, we first tried to take aged mice (fifteen months old) and replace their bone marrow cells with cells from young mice. Similar to bone marrow transplantation performed in leukemia patients, the recipient immune system should be first destroyed by irradiation to create vacancy for the donor's bone marrow cells to replace them; without vacancy there is no room for the donor cells. At first we were disappointed to find out that it hadn't worked: The aged mice that got the young bone marrow transplant were still cognitively impaired. But the biggest surprise lay ahead.

When we irradiated old mice and this time gave them a bone marrow transplant from age-matched donors—that is, from other old mice—these mice showed significant cognitive

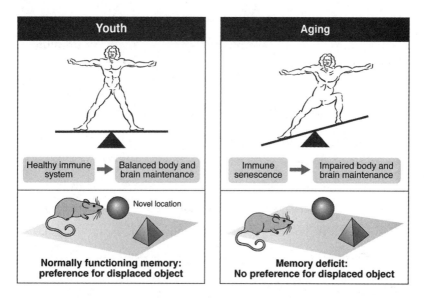

Figure 6. An aging immune system leads to cognitive decline. Animals are trained to become familiar with the arena in which they are placed, and in which two different objects are located in two corners. After a while one object is displaced and the animals are tested for the time they spend around the displaced object (novel location). Artificial aging of the immune system led to a reduced memory skill, manifested by the lack of preference for spending more time around the displaced object, indicating impaired memory regarding the primary location of the objects.

improvement. They performed the cognitive tasks almost as well as the young animals did.[18]

This result completely astonished us. We thought that the young bone marrow would prove to be the answer, but we found instead that it was the old bone marrow that had the properties we desired. We then understood what we had not initially considered. If we replaced old immune cells with young immune cells, we might lose some "memory immune cells" that were present in the old immune system. These are experienced immune cells that were educated throughout life by encounters with target antigens to recognize specific brain compounds, and that reside in the reservoir of the bone marrow, ready to act. (Note: Immunologists adopted the term "memory" to describe immune cells that have previously encountered their target compounds [antigens], and are therefore more responsive compared to naïve immune cells when encountering the same compound for the second time, or when encountering a related compound. Accordingly, immunological memory should not be confused with cognitive memories created and stored by the brain.) We found that indeed, more "memory T cells" could be found in the bone marrow of aged mice than in young ones with less immune experience. But if aged mice have more memory immune cells than young animals, why do aged mice experience cognitive impairment? They seem to have the necessary immune cells in their bone marrow but are somehow unable to use them to maintain their cognitive functions. We speculated that this was because there are too few of these memory immune cells in the circulation; even more critically, as the mice age, accumulated suppressor cells might inhibit function of memory immune cells, such as the "immune cells of wisdom" (figure 7). Indeed, we found that by eradicating the immune cells of the aged mice via irradiation we created a vacancy that allowed the transplanted old (and apparently "wise," including the repertoire of the experienced "memory T cells") bone marrow to repopulate the aged, newly "vacant," mice. We basically created a temporary space in the aged recipient mice, in which the donor memory immune cells could proliferate and renew on the expanse of naïve immune cells that had not yet met their antigen. Thus our manipulation of the aged mice broke the immune suppression and allowed expansion of the aged memory immune cells. In so doing we were able to reverse some aspects of the cognitive decline associated with aging and to rejuvenate the brain. This further supports our contention that the immune cells of wisdom are needed to prevent brain deterioration with age.[19] When we are young, we obviously have fewer of these immune cells of wisdom, but the brain does not need many of them, as wear and tear becomes an issue only with aging. When we are young there seems to be compatibility between the brain's needs and the immune system's ability to meet these needs. This is also the optimal time to boost your immunity as a hedge against aging "rainy days."

We still faced the most mysterious enigmatic issue: how immune cells, the cells that reside outside the brain—the immune cells of wisdom—affect the brain, and what can we do to access more of them as we age. In an experiment conducted in 2013 by Gilad Kunis and Kuti Baruch, graduate students on my team, we found that immune cells communicate with the brain from a remote region known as the brain's choroid plexus, a site located at the border of the brain's territory: the nexus between the brain and the circulating blood supply.

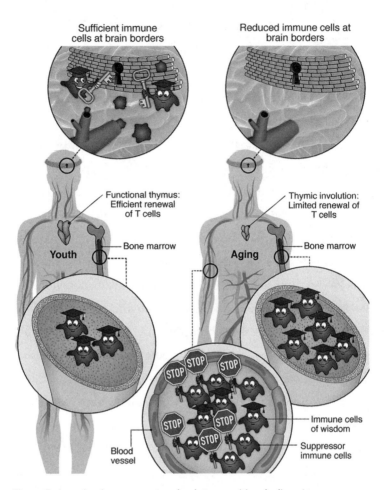

Figure 7. An aging immune system leads to cognitive decline. At a young age your immune system is at its optimal function: your thymus keeps renewing T cells, and "immune cells of wisdom" build up in your body and are recruited to your brain borders, where they assist your brain. As you age, your thymus shrinks, resulting in a limited renewal of T cells. "Memory immune cells," among which are the immune cells of wisdom, accumulate in your immune organs, even more than at a younger age, but are restricted by suppressor immune cells. This leads to reduced patrolling of immune cells at your brain borders, at a time when these cells are mostly required to cope with the wear and tear of the aging brain.

This site is enriched with blood vessels and is covered by a type of sheath (epithelial membrane) that surrounds many internal tissues of the body, including the lungs and the inside of our intestines. The choroid plexus acts as a command center for immune cells, some of which control brain functions remotely, outside of the neuronal tissue, by delivering molecules through the membrane.[20]

In fact, as you will see in the chapters to come, the choroid plexus is a unique interface between the brain and the immune system and a gateway within the central nervous system territory for immune cells. In addition to its role as a remote platform for border patrol by immune cells that affect the brain from afar, it also controls the access of immune cells to the brain tissue itself and prepares cells that are "on call" to enter and provide help to the tissue. In a study conducted in a model of spinal cord injury, led by a former graduate student in my laboratory, Ravid Shechter, we showed that following damage within the central nervous system, this unique interface between the brain and the circulation receives signals from the site of the damage and subsequently opens a biochemical route across the border, allowing selected healing immune cells that reside there to enter.[21]

Thus the choroid plexus serves major roles to keep lifelong brain function under control. First, it performs its long-recognized role of filtering from the blood the compounds needed to keep the brain healthy, thereby creating the cerebrospinal fluid. Second, it serves as an active immunological organ that senses the brain's needs and, with the assistance of the immune cells that reside in it, produces compounds that support the brain's health and activity from afar. Finally, it serves as a gateway for selected entering immune cells that support central nervous system repair. Unfortunately, this gate is subjected to changes in aging that reflect alterations in both the brain and the immune system, and that may interfere with signaling from inside and outside the brain. Only if the communication through this interface remains intact—which requires adequate signals from the brain that call for help and sufficient response by the immune system—will this interface continue to function as a gatekeeper of the brain. Once this communication breaks down, dormant diseases or brain aging may emerge or be manifested.

But how does the aging immune system limit the ability of the choroid plexus, the border that allows brain-immune communication, to support cognition? Are these changes reversible? With my students Kuti Baruch and Aleksandra Deczkowska, we have recently found that, during aging, the membrane of the choroid plexus undergoes immunological changes and develops pathological inflammation that disrupts its structure and reduces its ability to operate efficiently, both as a gate and as a manufacturer of compounds that nurture the brain. This blocks the necessary immune cells from entering the brain. We knew that a similar inflammatory process occurs in the airway filters of the lung in asthma and cystic fibrosis, decreasing barrier functions as well. These aging-induced changes impair the gate's ability to maintain daily brain functions and assist in repairing the damaged brain. By the bone marrow transplantation that we described, which involved resupply of the aged mice with experienced T cells, we were able first to break the general suppression of the immune system that develops with aging, and that restricts the function of such "wise" cells, thereby allowing the injected experienced T cells to prosper. Secondly, by reinforcing these immune cells of wisdom, we were able to restore the efficient functioning of the interface between the brain and the immune system. As we'd shown, this is crucial if we hope to reverse aging in the brain and restore cognitive function to youthful levels (see figure 7).[22]

Recently, in a study published in *Science,* we found that aged mice express a unique set of proteins (notably Interferon-ß), inhibiting immune communication at the brain's border. This set of proteins is also produced in the brains of healthy elderly humans. To determine what causes the aging brain border to produce Interferon-ß, we removed from young mice the cells that constitute this specific border and exposed these cells to fluid from the brain of aged mice in the laboratory. In response to the fluid derived from the aged brain, the young cells of the brain's border showed production of Interferon-ß and related proteins. This indicates that brain signals are responsible for this type of impairment in immune communication at the border of the aged brain.

But do these changes at the brain's border cause the loss

of learning and memory experienced by the aging population? To test this, we again employed the novel location recognition test. We took a group of aged mice that had defective memory and thus no preference for the displaced object (see figure 6) and injected half of them, directly into the brain's fluid, with a compound that blocks Interferon-ß. Two weeks after the injection, we retested the mice for their cognitive performance and found that the old mice in which Interferon-ß was blocked at the brain's border showed improved memory. They also had greater numbers of new neurons, which generally decrease in old age. Accordingly, our findings indicated that by restoring normal immune communication at the brain's border of aged mice, we could reverse age-related cognitive impairments.

Thus we view the brain's border at the choroid plexus as an active display window to the brain; when something goes wrong inside our brain, it is immediately apparent by the behavior of immune cells at this border. When immune communication is suppressed, the brain's function goes awry. The important conclusion is that targeting the brain itself is not necessarily the sole way to fight its age-related deterioration; manipulating immunological communication at the brain's border may serve as an alternative and less invasive therapeutic strategy for restoring brain function. In other words, we can restore cognition by manipulating the immune cells in the interface at which the body meets the mind.[23]

Tony Wyss-Coray and his group at Stanford University

found that when young mice were exposed to the blood of aged mice, the young animals showed cognitive impairment. In our recent work, in collaboration with Wyss-Coray and his group, we found that in the young mice that had received the blood of old mice, the choroid plexus border showed elevated production of a protein found to impair cognition. It is thus possible that the immune dysregulation that develops in aged mice was transferred to the young animals, interfering with the normal control of the brain's border by the immune system. It seems that immune communication at the brain's border is influenced by different cues and signals from both the aged brain and the blood. These findings reinforce the importance of the immune system in maintaining cognitive performance, and in preserving this performance throughout life. [24]

CAN BOOSTING YOUR IMMUNE FITNESS MAKE YOUR BRAIN YOUNGER?

In 2011, when I first lectured about the idea that we could rejuvenate the mind by boosting the immune system, I was bombarded with questions and pleas from the audience: *Does what I eat affect my immune system and thereby shape my brain? Can supplements help? Does exercise keep the immune system young? If so, which kind? What can we do? Give us the magic recipe for maintaining a healthy mind!*

Every self-respecting lifestyle magazine has tips for aging gracefully; these include regular physical activity and maintaining a healthy diet, among others. Numerous studies have shown that these factors indeed alleviate symptoms of brain aging.

How can the workout in the gym or the vegetable salad we are urged to eat save our brain from age-related deterioration? Our observations suggested that the immune system might be the missing link between a healthy lifestyle and maintaining the function of our brain.

Studies have shown that (nonobsessive!) physical activity enhances immune function, specifically by boosting T cells. In a similar manner, restricting calories while maintaining a balanced diet delays aging of immune cells. Thus a healthy lifestyle helps preserve our brain functions, at least in part, by strengthening our aging immune system.[25]

Our view is that brain aging does not necessarily reflect chronological age, but is rather highly dependent on a person's "immunological age." Along with other physical, environmental, nutritional, and genetic factors, the immunological fitness of the individual will determine the progression and extent of brain deterioration as part of aging (figure 8).

This might partly explain why some people "age better" than others. As we age, the risk factors for brain senescence accumulate within our brain. However, the fortunate ones among us, who have strong and well-functioning immune systems, will be able to contain these risks and maintain normal brain functions. For those of us who are less fortunate, an intervention in the form of boosting the immune system might improve cognitive function and delay brain aging.

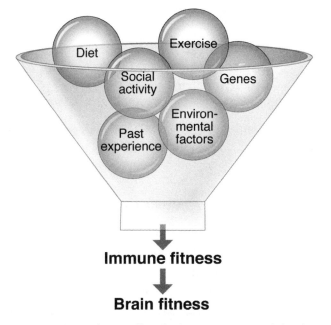

Immune fitness

Brain fitness

Figure 8. Various factors affect the immune system and thereby affect the way the brain ages and functions. Factors such as genetic background, past experience, social interactions, environment, diet, and exercise all act in concert to shape one's immune fitness, thereby affecting brain fitness.

Based on our research, if we fast-forward into the future, we envision elderly patients going annually to the gerontologist for an immune-boosting shot to keep their minds fit. Searching for the "fountain of youth" has led us to discover the immune system as the "elixir of life." We are currently working on developing such an antiaging immune-boosting treatment for restoring brain-immune communication at the borders of the brain.

KEEPING YOUR BRAIN SHARP
IN YOUR SIXTIES AND BEYOND

In addition to eating more of the immune boosting foods listed in the sidebar in Chapter 3, you can try these ways to keep your mind fit as you age.

1. Meditate

Meditation is receiving increased attention in the scientific literature. Several studies have shown that meditation improves cognitive functions and involves actual changes in different measures of brain activation, especially in areas linked to attention, learning and memory, and conscious perception. A few studies have shown that meditation can boost the immune system, increasing the levels of antibodies in response to pathogen exposure (figure 9).*

2. Exercise

Physical activity improves cognitive functions and decreases the risk of depression, dementia, and Alzheimer's disease. As we have seen, exercise has the capacity to enhance our immune system. You don't necessarily have to turn into a gym rat. Walking, swimming, even gardening are all excellent forms of exercise.†

3. Maintain Social Ties

Especially important for widows and widowers and elders without extended family. Volunteering can be a way to expand a social network.

4. Learn New Things

Attend free lectures at a local university or library, or master a new hobby, ideally one that involves exercise and expanding your social network. Many cities offer adult-education classes at low cost.

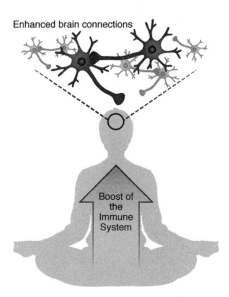

Figure 9. Meditation boosts the immune system, which in turn enhances brain connections.

5. Train Your Brain

Crosswords, Sudoku, and other puzzles can keep your brain sharp; even reading can help. Studies have shown that people who practice "brain training" improve or maintain their ability to perform daily chores as they age. Brain challenges such as reading, playing board games, and playing musical instruments are all associated with a reduced risk of dementia.[‡]

NOTES:

[*] Newberg et al., "Meditation Effects on Cognitive Function and Cerebral Blood Flow"; Cahn and Polich, "Meditation States and Traits"; Zeidan et al., "Mindfulness Meditation Improves Cognition"; Davidson et al., "Alterations in Brain and Immune Function Produced by Mindfulness Meditation"; "Meditation Improves the Immune System."

[†] Geda et al., "Physical Exercise, Aging, and Mild Cognitive Impairment."

[‡] Thompson, "'Brain Training' Benefits Seen 10 Years Later in Elderly"; Verghese et al., "Leisure Activities and the Risk of Dementia in the Elderly."

3

Stress and Depression

We have seen that the immune system controls our cognitive function and supports the formation of new neurons. These two aspects of brain plasticity allow us to adapt to our constantly changing environment. My team wondered whether the immune system also helps us to adapt to changes created by virtual enemies, the stressors we encounter daily throughout our lives. Could it save us from anxiety and depression?

How many of us occasionally feel overwhelmed by stress, pressured by our many responsibilities at work, school, or home? Perhaps your day begins with a hellish commute to work, continues with conflict with a boss or coworker, and ends in another traffic jam on the way back home. Maybe you wake up, heart pounding, convinced that you blew a crucial exam or job interview. Or at home, a squabble with your spouse flares up into a heated argument about money or parenting.

Yet most of us manage to cope with everyday stress. We make it through a rotten day optimistic that tomorrow will be better. We may first have to rant about our day to our near and dear. Then we unwind, everyone in his or her own way. We go to the gym to hit the punching bag or practice yoga. We do a crossword or escape into an online game. Or perhaps we head to a restaurant with friends or take the kids to a local park.

When we can't turn off the stress response, pathologies like depression and chronic anxiety can result, taking a heavy toll on our physical and mental health. What if the immune system is the gatekeeper of our mind and under certain stress conditions could help restore our brains to equilibrium and prevent these pathologies from developing? The work of my group in the past decade suggests that the immune system can do just that, tackling stress as though it were a pathogen or a dangerous foreign threat. But in order to understand whether the immune system controls our mental stability, we first have to turn some conventional wisdom on its head.

Anxiety and depression aren't microorganisms that physically invade and attack the body. How can the immune system fight against what seem to be virtual enemies? In this chapter, we will introduce the role the immune system plays in mood disorders and suggest ways to harness immunity to fight against the mental consequences of acute as well as chronic stress, and mood disorders. By emphasizing the immense contribution of the immune system, we don't mean to deny the value of classical antidepressants in restoring the chemical balance to the brain or the importance of talk therapy or behavioral therapy in helping patients understand their behavior and learn to control it. However, the strength of your immune system determines how you recover from traumatic events without developing posttraumatic stress disorder (PTSD), and to what extent your vulnerable mind is protected from anxiety and depression. By boosting your immunity, you can protect your mind against mental dysfunction.

In response to a perceived threat, our body is flooded with hormones, priming it to react. We become hyperaware, alert, and on guard. We're primed to battle or to run away. This "fight or flight" response was first described in 1915 by the American physiologist Walter Bradford Cannon. Also known as hyperarousal, in the animal kingdom this is the classic response of prey animals to predators. In humans' distant evolutionary past, the threat might have been a lion's roar or a crack of thunder. Now it's more likely to be an alarm or a siren. Your sympathetic nervous system immediately goes into high alert. This is the part of your autonomic system that increases your awareness, priming you to act. It releases a cascade of hormones to prepare your body to fight or flee. Your body secretes catecholamines, including the major stress hormones adrenaline and noradrenaline. These hormones increase your blood flow and make your heart pound. They are what make your hair stand on end, presumably to make you look bigger and more intimidating, just as the hair on a cat bristles when it arches its back. The sympathetic nervous system also dilates your pupils, sharpening your vision so you can see danger approaching. This hormone cascade allows humans to adapt almost instantly to new circumstances or to respond to a threat.

At the same time, the hyperarousal response also shuts down another system, the parasympathetic system. This is the system that calms and relaxes us. This system is responsible for the "rest and digest" response. As digestion shuts down, energy, oxygen, and nutrients are diverted from the stomach and delivered instead to the brain and muscles. After all, you don't need to digest a three-course meal while running away from a predator. These changes that occur in our body in response to stress, collectively defined as allostasis, are essential to our survival. They help us adapt to a changing and unpredictable world. We can view allostasis as a mechanism that creates a new state of physiological stability.[1]

In the animal world, the prey animal either fights the predator and wins, or runs away and escapes; otherwise it will not survive. In either case, the source of the stress rapidly vanishes. Robert M. Sapolsky, a professor of neuroscience at Stanford University, suggests that animals such as zebras experience only short bursts of stress and are therefore less susceptible to stress-related diseases. A zebra is either eaten by a lion, or escapes and survives. If it escapes, the zebra's physiology quickly returns to equilibrium, and it continues its normal placid life, grazing on the African plains. However, for humans, life is not so simple. Life is often full of worries, some of them groundless: as Mark Twain said, "I am an old man and have known a great many troubles, but most of them have never happened."[2]

In modern life, the stress rarely lets up. The body's attempt to adapt to the ongoing stressful condition is counterproductive. Thus, under the burden of seemingly unending stress—unemployment, major illness, caring for a sick child or a parent with Alzheimer's disease—some of us will fall into despair. This is *chronic* stress, which unless relieved can have a host of ill effects on mental and physical health and lead to dangerous self-medication through alcohol, drugs, or other risky behaviors, such as gambling, compulsive spending, or sex addiction.

Continuous stress thus causes wear of both psyche and body. The neuroscientist Bruce McEwen calls it "allostatic overload." Such overload causes pathology throughout the body, but the pathogen isn't a virus or bacterium. It's our body remaining on high alert even when there is no longer any real danger.

These pathologies in the body include elevated blood sugar, heart palpitations or panic attacks, and suppressed immunity. If not corrected, they can lead to gastrointestinal problems, asthma, and diabetes. This concept was introduced by Hans

Selye, an endocrinologist who performed pioneering studies that revealed how stress affects the human body. Selye showed in his experiments that many illnesses, such as the ones we mentioned above, could be linked to continuous stress that led to an overproduction of stress hormones. Selye coined the term "general adaptation syndrome" to describe a condition under which the body's first response to any kind of stress is an alarm reaction that involves release of stress hormones, which prepare the body to the immediate response to the stressor. If the stress continues, the body tries to adapt to the new situation, resisting the stressor by making the required changes. Normally the stress symptoms vanish at this stage. However, if stress persists, the body's resistance energy could eventually run out, resulting in the body's exhaustion. Such exhaustion is manifested by the loss of function of vital organs, causing disease and, if the stress continues, even death.

Continuous stress also affects the brain, disrupting its chemical reactions. Whether the threat is real or virtual, a creation of the brain itself—"a worry," as Dan Zadra puts it, that is "a misuse of the imagination"—the brain reacts powerfully. It is as if the brain is being held captive by stress, forced to impose changes and adapt. When the stressor is transient, these changes help your brain regain its biochemical balance and withstand the stressful event. However, when the stressor is continuous, your brain fails to restore its chemical equilibrium, which might culminate in a range of mental disorders, including anxiety, depression, and PTSD.[3]

More puzzling is why some of us recover from these episodes of stress without any mental symptoms while others, upon experiencing the same traumatic episode, develop anxiety and even PTSD. Clearly, the cognitive memory of the event is similar or even identical, but the behavioral symptoms this memory creates are totally different. Does it relate to the capacity of our immune system to restore the brain's chemical equilibrium and to erase the biochemical signature induced by the stress? If so, how?

THE BLACK DOG: DEPRESSION

In 1990 William Styron, the novelist best known for *Sophie's Choice,* published a memoir about his struggle with clinical depression, a struggle that brought him to the brink of suicide. The book, *Darkness Visible,* shone a bright light on an illness whose sufferers had long lived in the shadows. Published at the outset of the Prozac revolution, the watershed in the way depression was perceived and treated, this book refuted the pervasive assumption that depression was a result of personal failure or a weak character.[4]

In the ancient world, melancholy or depression was seen as a purely mental problem, the result of demonic possession or evil spirits. Hippocrates was the first to suggest a physical mechanism for depression in the form of imbalanced bodily fluids, or humors. This tug of war over the origin and nature of depression persisted for centuries, with first one and then the other view gaining ascendancy.

With the dawn of the twentieth century, advances in psychology, led by the works of Sigmund Freud, Erik Erikson, and Carl Jung, supported the view of depression and many other behavioral disorders as mental problems. Freud claimed that depression derived from emotional causes rather than physical ones. He linked melancholy to mourning, either over a real loss (such as the death of a loved one) or a symbolic loss (such as the failure to fulfill one's ambitions). He advocated the use of psychoanalysis to relieve depression.

About the same time, an opposing approach emerged, headed by the German psychiatrist Emil Kraepelin, who believed that each of the psychiatric disorders had an underlying biological and genetic cause.

As with many other important medical discoveries, the first class of antidepressants was discovered quite by accident. In the early 1950s, doctors who treated tuberculosis patients with a new drug, iproniazid (trade name: Marsilid) noticed that their patients became extremely happy. At the other end of the spectrum, patients treated with a drug to control their blood pressure (reserpine, trade name: Raudixin) complained of "feeling blue" and "going insane," with one patient attempting suicide. Scientists were puzzled by the profound opposite effects of these supposedly unrelated drugs on the moods of their patients.

The 1960s brought enormous advances in neuroscience. By the end of the decade, researchers and clinicians alike had realized that nerve cells in the brain used chemicals to talk to one another. These chemicals were neurotransmitters. Scientists found that the newly discovered euphoric and depressive drugs (iproniazid and reserpine, respectively) contained chemicals that interfered with the normal "chatting" among nerve cells in the brain. After this discovery, the perception of clinical depression and its treatment underwent a profound transformation. Previously seen as a sign of weak mind, and almost a personal failing, depression began to be accepted as a disease that resulted when chemicals in the brain were out of balance. It was a short step to treating depression with drugs that restored that chemical balance. Pharmaceutical companies were soon in the race to develop and patent such drugs.

At the turn of the twenty-first century, researchers discovered that antidepressants might do more than "simply" adjust the levels of certain chemicals in our brain. These drugs actually support the formation of new neurons in the brain, reorganize the circuitry of nerve cells in the brain, and increase neurons' adaptability, all of which are essential features of brain plasticity that are influenced by the immune system, as we discussed in the previous chapter.[5]

The question remains: How do some of us experience a given situation without any symptoms of anxiety while others experience unbearable mental stress? Is it just the genetic hand we are dealt at birth? Could it be the culture and family environment in which we are raised? Is it what we eat and drink? Or the fact that we exercise, talk with a close friend, or take a walk in the park to beat our blue mood? In fact, as we will show, our immune system is key to our ability to fight the virtual enemy of chronic stress; the way the immune system works determines the consequences.

It is important to note that the term "depression" represents various symptoms, traits, syndromes, and diseases that may have distinctive origins and causes. The experiments described in this chapter are based on animal models, which mimic symptoms characteristic of several depressive disorders, and are not intended to represent a specific type of depressive disorder diagnosed in humans. Having said that, I strongly believe that the immune system is a key factor in most depressive ailments, and therefore manipulation of the immune system, as indicated from our results, has the potential to assist at least in some of these ailments.

CAN THE IMMUNE SYSTEM FIGHT STRESS AND DEPRESSION?

A little more than a decade ago, in 2004, in one of my nonstop discussions with my graduate student at that time, Jonathan Kipnis, we were tossing around the idea that perhaps *virtual* enemies, such as chronic stress, might also be handled by the immune system. This was a few years after we had made the discovery, in 1998, that the immune system helps us to fight against injuries to the central nervous system, which at that time was a big surprise.

Jonathan and I wondered whether our brain is assisted by the immune system to cope with stress, and if so, whether there is a way to reduce patients' dependence on antidepressants by marshaling the immune system against the virtual predator—dealing with the biochemical consequences of chronic stress. Living life to the fullest is a painful enterprise. There is no way to totally shield ourselves against life's sorrows. We wondered whether there is a way to harness our immune system to protect us from developing helplessness and depression.

Could the same system that fights off and kills bacteria and viruses have an effect on conditions that involve not pathogens but rather virtual enemies like anxiety and depression? If so, how? We reasoned that if we could understand how the immune system helps the brain restore its balance after a painful mental episode, and what goes awry in the brain when the immune system malfunctions, perhaps we could develop a way to overcome the malfunctions, boost those immune mechanisms, and make minds more resilient.

Before approaching these questions we had to understand the way the immune system functions under stressful conditions.

THE IMMUNE RESPONSE TO ACUTE STRESS

You often encounter the cliché in the health care press that you feel bad and you develop diseases because you are under stress. While continuous stress is one of our biggest enemies, short-term stress can actually boost our immunity, reduce our fear, and help us to optimize our physical and mental performance—as long as it doesn't last too long. So perhaps if we understand how the immune system helps us to overcome the consequences of acute stress, we will have a better idea of how chronic stress undermines immune system activity and eventually results in depression. In the process, we should discover how to modify immune activity as a way to treat depression.

I recall my own personal experience that I share when teaching students, as a way of showing them stress as a positive mechanism. On the day that I had to give my defense lecture for tenure at the Weizmann Institute, five minutes before the lecture, I went to the toilet in the basement of our building, rather than in our department, which was located on the fourth floor. I was locked in; unable to unlock the stall door. Without even thinking, I found myself climbing on the handle of the door, and jumping across the door in the small space between the six-foot-high door and the ceiling above. There was no way on earth that I could have accomplished such an athletic feat without the adrenaline surge of my stress!

Although we often link stress to the onset of devastating diseases, under the belief that stress suppresses our immune system, in reality, short-term stress can actually boost immunity. Firdaus Dhabhar and Bruce McEwen were the first to show that short-term stressors activate immune cells, summoning them to essential or vulnerable areas where they begin to act. Short-term stress that mobilizes and redistributes immune cells in the body may accelerate recovery from surgery and boost the effect of cancer therapy or vaccinations. In fact, Dhabhar himself admitted, in his 2013 TED talk, that he happily chases his own children about the playground before they get their vaccinations in order to turn on the fight-or-flight response and rev up their immune systems.[6]

A transient boost to the immune response under acute stress makes sense in terms of evolution. Going back to the zebra example, what would be the point of allowing the zebra to deal with its stressor (escaping the lion), if it was then left with a suppressed immune system, resulting in an impaired ability to repair the wound it acquired during its escape? Following this line of thought, we speculated that, on top of its recognized function of protection against infection, the immune system also helps the zebra restore its brain chemistry to a balanced state, enabling it to react appropriately, and to proportionally discriminate between a real threat and a fictitious one, upon the next stressful event. Thus our zebra would be ready to run away from the next hungry lion, but at the same time it would avoid wasting energy running from a napping lion.

To test our hypothesis that the immune system equips not only the body but also the brain with a protective poststress mechanism, we exposed mice with and without working immune systems to a predator odor (in this case, cat urine) as a source of stress. The mice were exposed to the cat urine only briefly, simulating exposure to a transient stressor. All mice expressed an identical immediate response to the predator odor, which indicated that the perception of the stressor is not dependent on the immune system. Subsequently, though, in the process of recovery from the stress, the immune-compromised mice showed higher anxiety and greater fear response compared with mice with working immune systems. We tested these mice in an elevated plus maze (figure 10), a device commonly used to investigate anxiety in mice and rats. In such a maze, high walls enclose two of the maze arms, while the other two are open, with a thirty-inch drop to the floor. Mice are curious creatures, and under normal conditions most will explore open spaces. But if they are anxious, they will usually seek an enclosed shelter and stay there. Following exposure to cat urine, immune-deficient mice spent less time exploring the open arms of the maze than their control counterparts with a working immune system. This behavior reminded us of the avoidance behavior that is a hallmark of PTSD. Patients with PTSD are anxious and numb, and generally avoid any places, events, or objects that remind them of the trauma. These results taught us that the immune system is involved in the ability to restore our mental equilibrium following short-term stress. Thus if we are immune suppressed and we are exposed to even the slightest transient stressful condition, we may develop long-lasting symptoms of anxiety.[7]

Mice with immune deficiency also show a stronger long-lasting startle response to sudden, loud noises compared with their immune-competent controls. We see this same response in the hyperarousal symptoms common among PTSD patients. In addition to startling easily, such patients can be prone to angry outbursts and suffer from insomnia. Overall, these experiments have taught us that the immune system is an impor-

Figure 10. Elevated plus maze, a setup commonly used to assess anxiety in mice and rats. Two arms of the maze are enclosed by high walls, while the other two are open. Under normal conditions mice will explore open spaces, but if they are anxious, they will usually seek an enclosed shelter and stay there. Following exposure to a smell of cat urine, immune-deficient mice spent less time exploring the open arms of the maze than their control counterparts with working immune systems.

tant partner in coping with acute stress, and without it, the victim of stress may develop long-lasting mental dysfunction in the form of PTSD.

In the search for the role of the immune system in coping with acute stress, we found that the stress signals to the immune system: "Help me!" In turn, there is an increased inspection of the brain by immune cells, which help the brain to restore the levels of certain compounds that are needed for regaining mental activity. One of these is the brain's major protective protein, brain-derived neurotrophic factor (BDNF), the absence of which scientists associate with reduction of mental sharpness and with depression.[8]

So a person with a well-functioning immune system will be better able to overcome the destructive biochemical consequences of acute stress that leads to long-term behavioral and mental disturbance. This individual will still remember the stressful episode, but his brain will regain its normal biochemical balance and normal functions, enabling him to respond appropriately to future stressful events. It's analogous to a homeowner turning off a burglar alarm but resetting the alarm so that it will go off in the event of a future break-in.

POSTTRAUMATIC STRESS—WHEN THE IMMUNE SYSTEM LOSES THE BATTLE

The hidden wounds of posttraumatic stress were first brought to public attention by war veterans. PTSD was diagnosed in 10 to 30 percent of military veterans who fought in Vietnam, the first Gulf War (Operation Desert Storm), and, more recently, Iraq and Afghanistan.

However, with about 7.7 million American adults affected by PTSD, combat experience is obviously not the only risk factor. PTSD can affect any individual who has suffered a traumatic incident himself or witnessed a terrifying event that occurred to his loved ones or even to strangers.

One of the key symptoms of PTSD is reliving the event.

This can include flashbacks, nightmares, and recalling upsetting memories. The body reacts as if the trauma has never ended, resulting in a continuous stress response that badly affects the immune system. Studies have shown that PTSD patients suffer from an impaired or unregulated immune response, characterized by reduced numbers of immune cells, altered gene expression of immune-related genes, and irregular levels of cytokines (molecules that facilitate immune-cell communication). A few studies showed that trauma can actually change the expression of immune-regulating genes, thereby dampening the immune system. This compromised immunity might explain why people with PTSD are more susceptible to diabetes and cardiovascular disease, among other disorders. Our studies suggest that such compromised immune response is also to blame for the failure of PTSD patients to cope with the ongoing psychological burden they experience. Promoting a balanced immune response might help these patients strengthen their physiological and psychological health.[9]

IMMUNE RESPONSE TO CHRONIC STRESS

As we have seen, both acute and transient stress induces an S.O.S. immune response that helps restore biochemical equilibrium in the brain, the very equilibrium that was perturbed by the stress. Only when the immune system is defective does transient stress result in long-lasting biochemical dysregulation that eventually culminates in chronic pathology such as PTSD. But what happens when stress is unrelenting? The continuous release of steroids triggered by chronic stress has been shown to disrupt the immune response, reducing the number and quality of protective immune cells that can be "called up" by the immune system. The ability of these cells to travel to any site within the body where they are needed is also impeded. Such a damping-down of the immune system makes us more susceptible to a range of infections and to cancer.[10]

In a study conducted by Dhabhar's group addressing such susceptibility, two groups of mice were exposed to ultraviolet (UV) radiation, which is a weak inducer of tumors. One of the groups was exposed to chronic stress, while the other group was left unstressed. In the stressed mice, UV-induced tumors appeared earlier. The stressed mice also had lower numbers of effector immune cells and cytokines, molecules that allow immune cells to communicate, compared with the unstressed mice. These results suggest that chronic stress suppresses the immune system, making the stressed mice more vulnerable to tumor growth. Another study, performed in humans, showed that participants who had experienced long-term stressful life events were more likely to catch cold after exposure to a virus. These studies and several others indicate that continuous stress dampens our immune system.[11]

But is the immune-compromised state imposed by continuous stress also to blame for the behavioral disorders? Does the lack of a strong immune system leave us more vulnerable to develop mental disorders like chronic anxiety or clinical depression?

Our experimental evidence suggests that, yes, a chronically suppressed immune response reduces our ability to handle stress. Our immune system is designed to help us fight against the biochemical consequences that stress and anxiety cause in our brain in a way similar to how it fights against pathogens. Almost paradoxically, a continuous source of stress suppresses the immune system at the time it is so desperately needed in order to fight the consequences of such stress. The good news is that we can take steps to overcome this compromised immune system and its devastating consequences. In other words, long-lasting chronic stress creates a vicious cycle by dampening the immune system, leaving us more vulnerable and less capable of withstanding further burden of stress. We had to come up with a solution that would break this negative loop— by boosting the immune system. Could we design a vaccine to protect the mind?

A VACCINE FOR THE MIND

The fact that mood disorders could be influenced by a malfunctioning immune system suggested to us that by restoring immune balance or even boosting it above normal levels, we might be able to prevent, alleviate, or even cure depression. We needed to find a way to overcome the immune system breakdown caused by continuous stress, or to boost our immunity as a treatment to avoid consequences of traumatic or continuous stress. Our studies have shown that this can be accomplished by boosting the response of the unique subset of immune cells, the "protective autoimmune cells," described for the first time by my team in 1999. These are white blood cells in the form of memory immune cells (T lymphocytes), which recognize the brain's proteins (for more details, see the Primer). We defined these cells in the previous chapter as "cells of wisdom," needed for lifelong cognitive performance. Could these same immune cells affect our mental stability?

In these experiments, laboratory rats were maintained under continuous conditions of mild stress and irritating environmental factors including exposure to flashes of light, white noise (which rats find stressful), dirty cages, and restricted food and water, which together induce behavior similar to that of a depressive state. How do you objectively measure depression in a rat, which can't describe its mental state to the experimenter? There are several ways to measure rats' depression, one of which is using sugar water. Stressed rats exhibited a core symptom of depression called anhedonia. From the same Greek word for delight that gives us the word "hedonism," anhedonia refers to the failure to obtain pleasure from things that were previously enjoyable. In the case of the depressed rats, their preference for sugar water was impaired.

Another way to measure depression is by placing the rats in a glass tank filled with water. The rats could not escape over the slick, glass walls of the tank. Stressed rats spent less time swimming before they stopped and appeared to give up and resign themselves to drowning (at which point we quickly removed them from the water).

To boost the levels of cells of wisdom, which recognize

the brain's proteins, Gil Lewitus (at that time a graduate student in my team, now a postdoctoral fellow at McGill University) vaccinated the rats with modified fragments of brain proteins, which could only weakly "turn on" the immune cells of wisdom, an approach intended to safely boost the levels of such cells without risking a reckless inflammation. To create such a safe vaccination we used a principle commonly used for developing vaccines against viruses. To make a vaccine against a virus, scientists use a weak or killed virus rather than the live virus, in order to get the benefit of a strong immune response without risking the disease that the live, whole virus can cause. In the same way, vaccination with a modified fragment of a brain protein can safely boost levels of protective autoimmune cells without the danger of provoking autoimmune disease. We found that indeed this simple vaccination boosted the protective autoimmune cells and was sufficient to prevent the rats' anhedonic state, the desperate behavior in the water tank, and other symptoms of anxiety. In addition, the vaccination restored the rats' levels of nerve-protective factors and promoted creation of new nerve cells. In this way, the vaccination breaks the vicious cycle in which depression leads to slowed creation of new neurons, which thereby further escalates the depressive conditions. Studies have shown that fluoxetine (trade name Prozac), the extremely popular drug for depression, known primarily for its effect in regulating the balance of serotonin, the brain's compound of happiness, also acts by promoting the brain's cell renewal, as does our vaccination effect. In other words, it seems that the immune response can naturally do what the Prozac medication does in an artificial way. Thus the same immune cells of wisdom not only support cognition (see Chapter 2) but also serve as "natural Prozac," if only we could find a way to sufficiently harness those cells when we need them.[12]

So why can't we spontaneously use our immune system to prevent the suffering of PTSD and depression? Paradoxically, chronic stress dampens the ability of the immune system to fight depression. The vaccination can be viewed as a way of overcoming this immune suppression. Such a self-remedy has the advantage of helping the body do what it does best, without the need for drugs. In contrast to most vaccines that you received in childhood to protect you against potential dangerous invaders such as aggressive viruses or bacteria, our vaccination to save the mind is a form of a therapy. It does not wipe out a foreign pathogen but instead deals with boosting immune response in order to defend against our internal enemies and restore our brain's equilibrium and function.

So how does the immune response work? The cognitive perception of the stressful events remains forever embedded in the mind. The immune response cannot erase our cognitive painful memory of trauma, but it corrects the devastating biochemical footprint created by this trauma in the brain, which, if it persists, often leads to an imbalance of communicating molecules in the brain, and thus to depression. Yet if the immune response is too vigorous or is not properly terminated, it may be counterproductive, leading to chronic inflammation, which further escalates depressive symptoms. So balanced immune response is the optimal answer for fighting depression!

How could the role of immunity in depression have escaped scientists' notice in the century since Freud? We now know that head-to-toe immune protection comes with its own side effects. As has been known for decades (and as you probably recall from the last time you hurt your knee or got a scratch), any healing process brings with it a transient phase of pain, swelling, heat, and redness, collectively called inflammation, sometimes accompanied by fever. All of these are signs that the body is attempting to heal itself by recruiting immune cells. Yet when these signs persist, a new disease emerges that is manifested by chronic inflammation. Unfortunately, in the context of depression, it happened that the bad side of inflammation was the first to be noticed, as often happens in medicine, leading to a negative perception of inflammation in this and other pathologies for many decades.

When we decided to study the role of immune cells in brain repair and encountered the negative reputation of immune cells in all brain pathologies, we went back to the early definition of inflammation by the Russian biologist Élie Metchnikoff. He referred to *physiological inflammation,* a process that restores harmony following any threatening condition. When this process is not switched off promptly when the threat is gone, problems arise. Instead of healing, the inflammation response develops into a chronic pathological condition. It is that *pathological* inflammation that creates a favorable environment in the body for many diseases that plague humanity, including cancer, diabetes, and fatal neurodegenerative diseases, such as Alzheimer's and ALS.

Our studies showed that the immune cells are essential if we are to cope with stress and prevent anxiety disorders and depression. We found that the *good* side of the immune response is crucial if we are to maintain a healthy mind. Given the mainstream perception of inflammation as pathological, even the suggestion in our theory that immune cells protect the brain went against the conventional scientific wisdom.[13]

It is the immune system that copes with the biochemical damage these virtual enemies create. It does not eliminate the stressor or run away from it, as in the animal kingdom, but rather neutralizes its bad biochemical consequences. Therefore, when the immune response is impaired, the biochemical damage is left unresolved. However, if the same immune response fails to end promptly, it may become the pathological inflammation that can make the damage worse. To achieve the optimal result the immune response must be well regulated, but not completely shut down.

So how can we build up the immune system as our "natural" stress resilience mechanism? A stressful episode sets off an alarm within the brain, starting a cascade of immunological events. It begins as an S.O.S. from the brain, calling for immune assistance from the blood, eventually culminating in boosting levels of "memory" immune cells specific to the brain, which remain in the body long after the initial episode is over. Throughout our lives, without our being aware of it, occasional mild and transient episodes of stress are capable of building up pools of experienced immune cells that will serve as an immunological memory. These are the immune cells of wisdom, which help the individual to cope better with the next stressful event. In contrast, chronic stress suppresses such an immune potential.

According to the "hygienic theory," raising a child in a sterile environment undermines the ability to build a strong immune system to fight bacteria; similarly, I propose that a regimen of "mental hygiene" that overprotects children from everyday difficulties creates mentally fragile kids. They grow up into adults who can't deal with stress. Accordingly, immunization with brain protein fragments can serve as a shortcut in priming the immune response, thereby increasing our resilience to stress without our having experienced a traumatic episode in the past.[14]

WHAT YOU CAN DO: BOOSTING IMMUNITY TO
COUNTERACT ANXIETY AND DEPRESSION

Now we can answer one of the questions with which we opened this chapter: Why are certain people more prone than others to develop stress-related psychological disorders and even to fall into depression? We now know that the immune system is a major component of our mental health, a built-in form of mental therapy. Our world is full of stress-

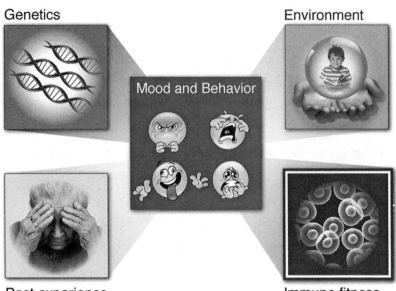

Genetics

Environment

Mood and Behavior

Past experience

Immune fitness

Figure 11. Immune fitness determines our mood and behavior. The immune system is a major component of our mental health: Our immune fitness dictates the way we cope with stress and contributes to our behavior, along with genetics, past experience, and the culture and environment.

ful events, and we all experience times of intense sorrow and anguish. But those of us who have stronger immune systems are more likely to cope with the stress, to overcome those bad days, and to continue functioning normally without developing mental aggravation (figure 11). Some of our ability to cope depends on nature and genetics: the strength of our immune system depends, at least in part, on our inherited genes. Studies by my team of researchers have shown that healthy mice with distinct genetic backgrounds (nature) but no differences in the way they were fed and handled (nurture) responded to stress in completely different ways. Some strains of mice recovered from the stress without showing behavioral malfunctions, whereas others developed signs of PTSD.[15]

Still, there are steps we can take. Attempts to design vaccination for stress-related psychological disorders are under way. Until those therapies are available, each of us can, by making lifestyle choices with immunity in mind, maintain our minds. By harnessing the power of your natural immune system, the body's natural Prozac, you can cope better with stress. You may even be able to delay or prevent depression and reduce your dependence on antidepressants. There's no doubt you will enhance your quality of life.

HELP FROM AFAR: HOW IMMUNE CELLS
HELP THE NEEDY BRAIN

If immune cells are excluded from the healthy brain, how can they help maintain our everyday resilience in the face of stress? Our studies showed that immune cells could do their job through a remote mechanism orchestrated from the brain's borders, especially from the blood–cerebrospinal fluid barrier located in a brain structure called the choroid plexus. The choroid plexus is located outside the brain's tissue, and consists of a layer of epithelial cells separating the brain from the blood circulation and receiving signals from both the nervous and circulatory systems. For decades scientists believed that the only role of this tissue was to produce the cerebrospinal fluid that circulates along the brain and spinal cord, and to enrich this fluid with only those factors and chemicals that are essential for brain function.

Through our work it became clear that the choroid plexus acts as a platform for circulating immune cells to communicate with the brain. When the brain tissue calls for immune assistance, the brain delivers a signal to this border and in response, the immune cells, located within this border, secrete molecules that can cross the border, support brain repair, and help restore the brain's chemical balance without actually entering the neural tissue. If this help is not enough, this border will then serve as an active gate, selecting and shaping the immune cells before they enter the central nervous system territory, much like soldiers are selected and trained at a frontline military base before they cross the border and enter a war zone. Vaccination with brain-related proteins helps boost the levels of protective immune cells at the brain's borders, and it appears to help summon such immune cells through the choroid plexus. We are currently conducting further research to address the role of the choroid plexus in promoting protective immunity in the brain. The brain is not the only organ that employs such a unique platform for interaction with the immune system. The eyes, the maternal-fetal interface in the placenta, and the testes all contain a similar system of epithelial-based gates that ensures optimal immune communication.[16] (For more about the choroid plexus, the blood–cerebrospinal fluid barrier, and the cerebrospinal fluid see the Primer.)

WHICH STEPS CAN YOU TAKE RIGHT NOW TO BOOST YOUR IMMUNE SYSTEM?

Though we are not a physical therapist or a dietician, we would like to provide some examples of how your choices of food and the way you choose to spend your leisure time can boost your immunity and thereby ward off anxiety and depression.

Many studies over the past ten years have revealed that diets rich in fruits, nuts, vegetables, fish, and spices may prevent cognitive decline and reduce the risk of developing mental disorders. These foods are rich in antioxidants, flavonoids, and certain vitamins known to boost immunity. Some research has shown that omega-3 fatty acids, a nutrient derived from seafood such as fatty fish and shellfish, canola and soy oils, and flaxseed, protect against cognitive decline in the elderly and supports the developing brain in the growing fetus. Eating a diet rich in these foods can be as simple as eating "lower on the food chain" a few days a week, and filling your plate with naturally colorful foods. Colorful fruits and vegetables are richest in antioxidants and other bioactive compounds.[17]

A study conducted at Ohio State University in 2011 suggests that omega-3 fatty acids can reduce anxiety among healthy young adults. A group of medical students received either omega-3 capsules or a placebo. Participants receiving the omega-3 pills showed a 20 percent reduction in anxiety symptoms. Blood samples taken from these students showed a significant reduction in molecules known to play a part in pathological inflammation. Omega-3 plays a role in strengthening the immune system. It activates or deactivates particular immune cells, helping the body switch off destructive inflammation on one hand, and boosting a protective immune response on the other.[18]

Another important immune booster affected by what we eat is vitamin D3, found in eggs and fatty fish, such as mackerel and salmon. Your body creates vitamin D3 after exposure to sunlight. An incomplete form of vitamin D3 is in your skin. Sunlight activates this incomplete form, and your liver and kidneys convert it to a final form your body can use. When you eat eggs or fatty fish, you gain benefit from the vitamin made in the body of the chicken or the salmon. (A different form, D2, is routinely added to milk in North America.) However, as most of us in the developed world spend our days indoors, an increasing proportion of the world's population suffers from vitamin D3 deficiency. Such deficiency is associated with mood disorders, anxiety, and depression. Vitamin D3 promotes a balanced protective immune response (figure 12).[19]

Similar evidence exists for regular exercise. Many of us feel that our morning run or the hour we spend in the gym helps us better cope with the stressful day to come and lifts our mood. There is now strong evidence that exercise helps us avoid depression, and that people already suffering from mild depression can use exercise to help restore a more positive

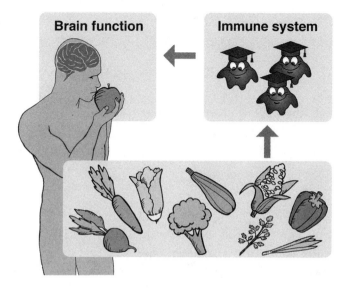

Figure 12. An apple a day keeps the (brain) doctor away. Immune-boosting foods support brain functions. Feeding your immune system a diet rich in fruits, nuts, vegetables, fish, and spices may prevent cognitive decline and reduce the risk of developing mental disorders.

outlook on life. As we discussed in the previous chapter, studies carried out by my team show that the benefits of physical activity on brain plasticity depend on a functioning immune system. These beneficial effects are diminished if our immune system is impaired. Other studies have shown that physical activity decreases risks of dementia and Alzheimer's disease and improves cognitive functions such as learning and memory.[20] These findings endorse the common recommendation to patients to replace their antidepressants with sport or other physical activities, to adopt the "brain fitness program." Exercise is vital for maintaining mental fitness, and it can reduce stress.

———∿∿∿∿∿∿———

TOP 10 IMMUNE-BOOSTING FOODS

Not sure where to begin? To start, try introducing into your diet one or two of these food items each week. Even if you already eat a varied diet full of fresh food, consider adding a few of these foods to the mix, if you don't already enjoy them:

Fish and shellfish (contain omega-3 fatty acids, selenium)

Plant oils such as olive and canola (omega-3 fatty acids)

Deep green and dark red/orange vegetables (beta-carotene and other vitamins), preferably raw or steamed to retain nutrients

Citrus fruits and berries (vitamins C and E)

Yogurt, kefir, or other dairy products containing live cultures (probiotics)

Oats and barley (beta-glucan)

Garlic (allicin)

Black or green tea (L-theanine; decaf tea also contains it)

Mushrooms, especially shiitake and varieties other than white "button" mushrooms

Nuts, seeds, and beans (omega-3 fatty acids, zinc, and selenium)

NOTE:
"Slideshow: 15 Immune Boosting Foods"; MacMillan and Schryver, "9 Power Foods That Boost Immunity."

———∿∿∿∿∿∿———

Even more than nutrition, exercise has the capacity to enhance the immune response. Scientists have not completely deciphered the mechanism, but they have some hypotheses. When we exercise, our heart rate rises as our body pumps blood to our muscles. This increased blood rate allows immune cells to circulate around the body more efficiently and may also speed renewal of other immune cells. Aerobic exercise reduces the levels of stress hormones like cortisol and adrenaline. This damping down of stress hormones may also boost the immune system. Another hypothesis is related to the exercise-induced release of endorphins, the natural painkillers that our brain produces during physical activity, which are also responsible for the euphoric feeling known as the runner's high. Endorphins are brain-derived communicating molecules that can act directly on the brain, but they have also been found to affect the immune system. Regardless of the specific mechanism, when exercise revs up our immune system, we are better able to fight off anxiety and depression.[21]

The specific exercise is less critical than duration and frequency. Of course, moderation is the key. Be careful not to overdo it. Too much exercise actually *decreases* the numbers of circulating immune cells and increases the levels of stress-related hormones. When exercise becomes an obsession or an addiction, it loses its benefits and can begin to harm mental health. According to the U.S. National Institutes of Health a moderate program might include bicycling a few times a week, walking daily for twenty to thirty minutes, working out at the gym every other day, or playing golf regularly. In this connection, it is interesting to note that meditation also improves the immune system and the mind.[22]

Two thousand years ago, the Roman poet Juvenal coined the phrase *mens sana in corpore sano:* a healthy mind in a healthy body. Now we understand that the immune system is the body's agent in lifting our mood. The fitness of our physical body, which is manifested by the strength of our immune system, affects our mind.

4

Of Mice and Superman

The Immune Pro—Spinal Cord Therapy

When Superman asks for your help, how can you refuse? In 2000 I was invited to the New Jersey mansion of actor Christopher Reeve. In a riding accident in 1995, Reeve had been thrown from his horse and had landed on his head. The catastrophic injury to his spinal cord at the base of his neck left him quadriplegic, unable to use his arms and legs. He could breathe only with the help of a respirator.

Desperate to find a cure for his paralysis, Reeve was excited to learn about our results and progress made by others. Together with Robin Williams and other friends from Hollywood, he had launched a foundation to raise substantial funds for research into new treatments for spinal cord injuries.

I was inspired to meet this proud, intelligent man. He was well versed in all the current medical literature on spinal cord injury. A committed advocate for spinal cord research, he urged scientists to join forces to find a cure. I emerged from our meeting highly encouraged, determined to push forward with my own research.

The human spinal cord presents an evolutionary and medical mystery. Although some species of fish and amphibians can rebuild their spinal cords after injury, for humans it seems impossible. Once severed, the human spinal cord cannot mend itself. Our bodies can mend broken bones, regrow parts of the liver, and even heal after heart surgery. Almost alone of the human body's tissues, the brain and the spinal cord resist healing.[1]

———⌇⌇⌇⌇⌇———

Why should humans have such a fragile spinal cord? Why haven't humans evolved ways to heal injured spinal cords, as some other vertebrates are able to do?

Since the 1960s, researchers have proposed various explanations for the spinal cord's inability to heal after being severed or crushed. According to one theory, severe spinal cord injuries immobilized our vertebrate ancestors, rendering them dependent on their community and unable to defend themselves against predators. Paralyzed animals had almost no chance of survival and passing their genes. Thus these animals were weeded out by natural selection, and a mechanism for spinal cord repair did not have a chance to evolve.

A different explanation is that any healing process exacts a price on the body. Immediately after we get a cut on our skin, a repair process begins. This process includes the inflammatory immune response, cleaning away damaged cell parts, as well as scarring that forms a temporary replacement of the damaged tissue and a scaffold for repair of the injured site. As part of this process some healthy cells are killed in order to save the tissue as a whole. That's fine for skin cells that are replaceable, but spinal cord neurons are not expendable. According to this theory, the cost to repair a mammalian spinal cord may exceed its benefit.

In addition, healing of severe injuries requires the help of circulating immune cells. The access of such cells is restricted in the central nervous system due to the barrier system that enables the brain to function in a stable, secure environment, unaffected by the continuous fluctuations in the blood.

Thus many scientists assumed that over the course of human evolution, the central nervous system had to compromise. To avoid the loss of irreplaceable neurons and their optimal balanced environment, the central nervous system has "given up" the opportunity to have a classical healing process. It was left with the ability to heal only mild injuries, therefore having limited capacity for spontaneous recovery.

———————~~~~~~~———————

Spinal cord injuries present a serious public health problem, estimated at almost 180,000 new cases each year worldwide with varying degrees of disability, 12,000 to 20,000 cases per year in the United States alone. Most spinal cord injuries require expensive, long-term care, exacting a heavy economic toll on families and health systems alike. Patients with spinal cord injury must adjust to drastically altered lives, often with markedly reduced sensation, mobility, and independence. It can be a struggle to return to work and pursuits enjoyed before the injury. Because spinal cord injury takes a heavy physical and emotional toll on patients, researchers are searching for new treatments and therapies to increase the healing potential of the central nervous system. They hope to halt the spiral of damage and tissue loss that can occur following injury, and to find new ways to get the spinal cord to heal itself.[2]

—— ~~~~~~~ ——

In spinal cord injury, which bodily functions are lost depends upon the location and severity of the break. A break in the spinal cord of the lower back (lumbar) leads to paralysis of the legs and problems with bladder control and sexual function. An injury to the spinal cord higher up in the trunk (thoracic) can affect breathing and use of the arms. A spinal cord that is severed higher up, in the area of the neck (cervical), results in loss of function from the neck down. All forms of paralysis can result in loss of sensation to temperature, touch, and pressure.

Secondary complications following spinal cord injury may also vary depending on where and how severe the injury is. These can include pneumonia, unstable blood pressure, irregular heartbeat and blood clots, exaggerated reflexes, muscle atrophy and spasms, loss of bladder and bowel control, chronic pain, and pressure sores.[3]

———∿∿∿∿∿∿∿∿∿———

Throughout history, spinal cord injury was considered a catastrophe from which there could be no recovery. The Edwin Smith papyrus, the source of much of what we know about the medicine of the ancient Egyptians, contains a description of a patient who suffered the fracture of a vertebra in the neck toward the back and had such symptoms as complete paralysis, lack of bladder control, and sexual dysfunction. The ancient case history concludes that this was "an ailment not to be treated."

Even in the early decades of the twentieth century, most people who suffered spinal cord injuries died within a few weeks. But by 1950, enormous advancements in neurology allowed more and more of these patients to survive for prolonged periods. Now the goal was no longer just keeping patients alive but restoring as much as possible their quality of life, and perhaps allowing them, one day, to regain some of the function the injury had stolen from them.

Despite these advances, mending vertebrae and offering patients physical therapy could not regrow a damaged cord. Thus the scientific community continued to devote research into how severed cords might heal, keeping in mind the dogmatic claim of the influential neuroscientist Ramón y Cajal that spinal cord nerves cannot regenerate. This assumption

OF MICE AND SUPERMAN

plunged the field of spinal cord research into a long dormancy.[4]

—————〰〰〰〰〰〰—————

To heal successfully, a severed spinal cord must regrow nerve axons and reconnect them to the existing neuronal network. This process rarely happens in a severed spinal cord. In fact, one of the most notorious episodes in the field of spinal cord injury research was work by the neurosurgeon Carl Kao. In the late 1970s Kao transplanted grafts of nerve segments from legs into the injured spinal cords of dogs, claiming to induce regeneration and restore mobility to the dogs. Kao went on to perform similar surgery in humans without publishing many of his results in refereed scientific journals, leading many in the scientific community to accuse him of conducting unethical experiments on desperate patients.[5]

In the early 1980s, two researchers from McGill University in Canada, Albert Aguayo and Peter Richardson, replaced part of an injured spinal cord with a segment of leg nerve (not part of the central nervous system). Such a segment could not function for conduction of nerve impulses but could potentially serve as a scaffold, supporting the growth of severed axons that normally fail to grow across the injury site. Using an enzyme derived from the horseradish plant to track the axons of neurons in the spinal cord, Aguayo and Richardson were able to show that the leg nerve transplants indeed served as growth-supporting scaffolds along which the injured spinal cord neurons could regrow. They concluded that the injured central nervous system microenvironment, as Ramón y Cajal had claimed a generation earlier, does not support regeneration. Kao's experiments, providing transplants of non–central nervous system tissues to promote regeneration, proved better grounded than had been assumed.[6]

After Aguayo and Richardson elegantly proved the concept behind Kao's experiments, researchers around the world tried to identify either the missing growth-supporting compounds or the existing growth inhibitors in the tissue surrounding the severed cord that were preventing nerve cells from regrowing and reconnecting. Martin Schwab, a Swiss neuroscientist, discovered a novel family of compounds, present in the myelin membrane that covers the spinal cord nerve axons. Called Nogo proteins, these compounds play a crucial role in inhibiting regeneration. Schwab and his colleagues also designed antibodies that worked to block the activity of the Nogo compounds. In mice and rats, the antibodies caused long axons to regenerate and thereby partly restored lost function. Schwab's team is still conducting studies of Nogo antibodies to see whether they can jump-start spinal cord repair in humans. Subsequent to the Nogo, additional growth-inhibitory compounds were identified.[7]

Since the 1980s researchers had believed that scarring presented yet another obstacle to spinal cord regeneration and healing. To make that assumption is to see the glass as half-empty; as I will demonstrate later, scarring is also essential for repair, and becomes a drawback only if it persists. The immune cells are the ones that keep scar formation under control.[8]

A BROKEN PROMISE: STEROIDS FOR SPINAL CORD INJURY

I first confronted the nearly complete inability of the spinal cord to heal itself in the summer of 1978. I had just finished my doctorate in immunology and I was trying to decide on a postdoc research project. I found myself wondering why the human body can heal all kinds of wounds to the skin, even burns, but not an injury to the spinal cord. Why should this be so? With my background in immunology, I thought I should be able to provide new insights into this long-standing puzzle. I started by studying the healing process in the fish central nervous system.

Unlike mammals, fish and amphibians are capable of re-generating their central nervous systems after injury. We were following the repair process in fish whose optic nerves (part of the central nervous system, discussed in the following chapter) had been crushed. When we transplanted a graft of regenerating fish optic nerve into the crushed optic nerve of a rabbit, some regeneration of the rabbit's optic nerve was observed. There was something in the environment of the fish optic nerve that promoted regeneration that was missing in the mammalian optic nerve. We were determined to find out what it was.[9]

Our studies revealed that the missing factor originated in the cells that surround the nerve, but we still had no clue as to what this special fish factor was. As I was aware that immune cells are key participants in healing a wound, I suspected that it might be related to the interaction between the nerve and the immune system. In mammals the central nervous system is largely secluded from the circulation and the immune system. Fish are much less complicated. They don't have the sophisticated blood-brain barrier system that mammals have. Thus their brains are much more accessible to immune cells. I speculated that this direct immune accessibility is one of the differences that make fish able to regenerate spinal tissue and humans and other mammals unable to. About the time that I was ready to test this theory, I uncovered a research paper that almost stopped me in my tracks. It appeared that someone else had already solved the problem, arriving at a totally opposite answer!

A medical doctor and neuroscientist at New York University in the United States, Wise Young (now at Rutgers), claimed that he had solved the mystery of why the spinal cord does not heal from traumatic injury: inflammation. Young was among the first who attributed a major negative role to the inflammatory process following spinal cord injury, and he proposed to treat patients with high-dose steroid infusions within eight hours of the injury in order to arrest inflammation. Aggressively shutting down the inflammation as soon as possible after injury, Young claimed, could give patients who might have been consigned to life as para- or quadriplegics hope of recovering from severe spinal cord injury.[10]

Soon after Young published his research, the steroid methylprednisolone (trade name Medrol) was approved in Canada, several countries in Western Europe, and most Far Eastern countries for treatment of spinal cord injury. It quickly became the standard treatment in the United States as well.[11]

At first, learning of Young's work nearly derailed my research plans. I have met him several times and appreciate his devotion to improving the lives of spinal cord injury patients. Back in the 1980s, most experts had given up on finding a cure for spinal cord injury, but Wise was one of the few who still believed spinal cords could heal and that research was needed to discover how. While we shared a passion for research that could benefit patients, I remained skeptical about the overall approach of shutting down the immune system to promote healing. I had already begun to conclude that inflammation, one of the body's immune "first responders" involved in healing, could also play a similar role in recovery from spinal cord injury.[12]

At that time, the medical community was desperate for a cure for spinal cord injury, and it soon became enchanted by the Medrol. Only a few years later were concerns raised regarding the effectiveness and safety of methylprednisolone, concerns that gradually diminished doctors' reliance on this drug to treat spinal cord injury. Numerous follow-up studies cast doubt on the use of Medrol and whether it could promote healing for spinal cord injuries. Patients on steroids did not regain function at higher rates than their counterparts that did not receive the treatment. Steroids, being immune suppressors, decrease a patient's ability to deal with life-threatening infections and other complications. By 2002 more and more experts throughout the world had discredited the clinical benefit of methylprednisolone.

In 2013, guidelines released by the Congress of Neurological Surgeons and the American Association of Neurological Surgeons clearly recommended against the use of methylprednisolone for the treatment of spinal cord injuries. These new understandings were the first crack in the wall of the accepted wisdom that addressed as pathological the entire immune response following spinal cord injury. If it is so bad for the function of the spinal cord, I wondered, why did suppressing it not yield better outcomes in patients?[13]

OVERTURNING THE INFLAMMATION DOGMA

Studies have shown that injury to the spinal cord evokes a chain reaction in which the dying neurons spill out their contents, thereby creating hazardous conditions that prohibit repair and accelerate loss. This toxic microenvironment causes additional cells to die, damaging neurons that were spared in the initial injury, in a phenomenon known as secondary degeneration. Such additional cell death results in further loss of function and impedes healing.[14]

During the golden age of the steroid treatment for spinal cord injury, inflammation was perceived as the villain. It was seen as a key obstacle to healing the spinal cord and blamed for further cell death and the secondary degeneration. One of the first questions my team asked was whether this "villain" was unique to the central nervous system. We scoured the medical literature, reading dozens of reports on how the healing process unfolded in wounds in the rest of the body. We discovered that inflammation, the process that is assumed to inhibit spinal cord repair, is accepted as an essential interim process in healing wounds throughout the rest of the body.[15]

When a child skins his knee, parents at once notice the swelling, redness, and heat surrounding the scrape. After a while, a scar forms. The parents reassure the child that it is part of the healing process and that it will all be forgotten soon enough. And they are right. Injury outside the brain and the spinal cord is followed by increased blood flow and massive recruitment of immune cells to the damaged tissue to assist in healing. This is why the area appears hot, red, and swollen. The scarring phenomenon is also common in any wound healing process and is assumed to serve as a temporary replacement of the damaged tissue and a scaffold for repair of the injured area.[16]

My team knew that scarring and inflammation take place in the injured spinal cord. So why couldn't the injured spinal cord benefit from these processes that appear essential to the repair of tissues elsewhere in the body? What have the scientists been missing?

We suggested that the answer lies in the regulation of the inflammatory process and the scarring, rather than in the mere occurrence of local inflammation and scar formation. We viewed inflammation in the injured spinal cord, like that in the rest of the body, as a necessary response, but one that in the spinal cord doesn't end when it should. Extrapolating from the way healing proceeds in other parts of the body, I reasoned that for efficient recovery in any tissue, scarring and the inflammation should be transient. In my view, they serve as an interim stage in the healing process. Stages of that process need to occur in order and in synchrony. The toxic waste surrounding the wound should be cleared away by immune cells, and scar tissue should form to allow reorganization of the injured site as a way of preparing the damaged tissue for repair. Subsequently, scar tissue should be torn down, and the inflammatory response should power down. Altogether, these events create favorable conditions for protecting spared neurons and for subsequently building up new, functional tissue. Each of these steps along the recovery process is essential, and each involves distinct subclasses of immune cells, which should be switched on and off in sequence and on schedule, so that the overall repair process can be accomplished.

In the case of injury in the spinal cord, or in the central nervous system in general, it appears that such recovery steps often spiral out of control, which results in exaggerated scarring, chronic inflammation, and weak tissue repair. So the question that emerges is how we can provide the spinal cord with the proper immune response at the right time following injury to avoid this spiral out of control that turns a positive response into a negative one.

CAN IMMUNE CELLS REPAIR THE SPINAL CORD?

We were looking for a way to boost a protective immune response: one that is strong enough to jump-start recovery but is also timed precisely. In 1998 my group was the first in the world to introduce into the injured central nervous system immune cells, long considered harmful to this tissue. We were, in fact, injecting macrophages, immune cells that are involved in wound healing. At this time, doctors were still prescribing the steroid Medrol for patients with spinal cord injuries. When we first proposed that boosting levels of macrophages, the immune cells that are primarily known as cleaners, acting as "garbage disposers," might be the answer to treating spinal cord injury, we were fiercely criticized by some of our colleagues in the field. They claimed that the site of injury was already full of immune cells and that steroids were the solution. From their point of view, by adding macrophages we would be adding fuel to the bonfire. We thought that by adding macrophages to the severed spinal cord, we would be bringing one type of fire to fight a bigger fire, in line with Hippocrates' ancient claim, "Give me the power to produce fever and I'll cure all disease." Among the scientific community that studied spinal cord injury at that time, there was no awareness that macrophages could play a wide range of roles in wound healing, not all of which contribute to the fire, the inflammatory response; some are actually active in stopping the fire.

I still remember the minute when my team came back from the quarters where the animals were housed with big smiles on their faces saying, "The animals are walking!" We hardly dared to believe our result. I brought my three eldest children to see the miracle. Everyone in the family shared our excitement. Our colleagues and wider society were a different story. It was discouraging to have colleagues, many of whom I respected, trying to prove again and again that inflammation is *always* bad, ignoring the role of inflammation as a physiological response essential for survival. Contrary to our opponents' view, our findings showed that the transplanted macrophages, incubated outside the animal in the presence of skin cells and then injected into the site of injury, did not contribute to the pathological inflammation but rather secreted molecules that can support tissue repair. When these macrophages were injected at a certain dosage, at a precise time and location, they significantly enhanced the rats' recovery.[17]

Encouraged by these results, we embarked on further research that would ultimately lead to a clinical trial of macrophage therapy for injury to the spinal cord. This therapy involved isolating macrophages from a patient's own blood and growing them in the laboratory in culture with the patient's own skin cells. The resulting macrophages were injected into the patient's spinal cord within the second week following the injury.[18]

A VISIT FROM SUPERMAN

In the late 1990s, Proneuron Biotechnologies was founded to translate our scientific results into treatments for spinal cord–injury patients. Researchers were working nonstop to set the conditions for the translation of the technology from animals to humans. With the advice of the U.S. Food and Drug Administration, the first small clinical trial was kicked off, with the goal of establishing whether the therapy was safe.[19]

In the summer of 2000 in Colorado, a traffic accident left eighteen-year-old Melissa Holley paralyzed from the chest down. Her father learned about our clinical trial on the Internet. Within days, Melissa and her father arrived in Israel, where she volunteered to be the world's first patient, knowing that the procedure had been tried only in the laboratory and only in rats. In the months to follow, she began to regain feeling and functions in her lower body.[20]

Subsequently, other patients with severe spinal cord injuries have been treated, and the treatment has spared several of them total paralysis. Of the eight patients participating, three recovered clinically significant neurological motor and sensory function. In addition, the macrophage therapy restored sexual function to patients who might otherwise never have been able to bear children.[21]

In 2003 Christopher Reeve visited Israel to learn more about the cutting-edge paralysis research conducted in the country. During his trip he visited the Weizmann Institute of Science, where he met my research team. We updated Reeve on the progress we had made in our research and in applying the treatment regimen to patients in the first clinical trial (figure 13). Reeve met the physicians who were involved in the clinical trial, as well as some of the patients. He was especially impressed by meeting one patient who had suffered a severed spinal cord. This patient had macrophage implantation, spent two years in aggressive rehabilitation, and was able to walk

Figure 13. Christopher Reeve meets Professor Michal Schwartz at the Weizmann Institute of Science, 2003. Courtesy of the Weizmann Institute of Science.

with the support of parallel bars. "That's totally amazing," Reeve commented during one of his interviews. "That's the most impressive example of recovery from spinal cord injury I've ever heard of. So that helped to renew my hope. It was very satisfying, very rewarding for me to see how much progress [Schwartz is] making."[22]

During this same trip to Israel, Reeve was interviewed on *Larry King Live*. He praised the attitude toward medical research in Israel, and the Israeli scientists he had met. Reeve expressed hope for close cooperation between research institutes such as the Weizmann Institute of Science in Israel and rehabilitation facilities in the United States.[23]

A MACROPHAGE TO CONTROL INFLAMMATION

At the same time as the clinical trial, we were moving forward with our laboratory research. We were working hard to unravel the mystery of how to help heal a spinal cord injury, seeking a way to identify all the players at the level of cells and tissues. By mapping out how they interacted, we hoped to identify the factor or factors limiting healing and to make our initial therapy even more effective.

My colleagues and I started to believe that the severed spinal cord has the same need for inflammation as any other tissue in the body when injured. For reasons that we were only starting to understand, the spinal cord injury becomes a chronic wound that fails to heal, characterized by excessive scarring, continuous inflammation, and weak tissue regeneration.

We concluded that the ideal therapy would be one that could orchestrate the entire immune response cascade. We suggested that this could be achieved by boosting the appropriate immune response. This approach would help dial down local inflammation, restrict the formation of scar tissue, halt the domino effect of cell death, and allow the spinal cord tissue to regenerate. Access to such a protective immune response is what the central nervous system has been missing.

As we viewed the immune system as the natural defense force of our body, we wondered whether macrophages, sim-

ilar to those that we ended up injecting directly into the injured spinal cord, were spontaneously recruited from the
blood to the spinal cord following injury. If they are recruited,
why doesn't the spinal cord heal itself? Is it because too few
of these macrophages are recruited? Or because they arrive
too late? Or because they are not the right type? Whatever
the case, how can we most effectively recruit the most helpful
cells? We still had many questions, and we were determined to
get answers.

New research techniques developed by others allowed
us to get a deeper look into the way macrophages respond
to injured spinal cords. One of these was green fluorescent
protein, isolated from bioluminescent jellyfish and corals in
the early 1960s. Scientists are able to use this fluorescent protein to label cells cultured from experimental animals or the
human body. We collaborated with Steffen Jung, a researcher
of immunology from the Weizmann Institute who genetically
engineered mice to express the green protein in their immune
cells. Because the cells glowed green under certain light, we
could see where in the body they had traveled.

Together with two graduate students in my laboratory at
that time, Ravid Shechter and Anat London, we used these
"green" mice and were able to identify the macrophages that
are recruited to the spinal cord from the blood only after injury. The blood macrophages begin life as blood cells called
monocytes, and in response to some chemicals released by the
wound they are recruited from the blood to the damaged tis-

sues, where they become macrophages. We thus termed these cells monocyte-derived macrophages, to distinguish them from the native immune cells that reside in the spinal cord and brain from birth.[24]

We designed experiments to selectively prevent accumulation of such blood macrophages in the site of injury. This resulted in lower ability of the cord to heal. On the other hand, when we increased the recruitment of such macrophages to the injured site, simply by injecting these cells into a mouse's veins, the animal experienced significant recovery. We were amazed by the results: spinally injured mice injected with this type of macrophage—the cells had not been grown in a culture or gone through any other external manipulation—were able to move their legs and walk. Meanwhile, mice that had received a placebo injection remained almost completely paralyzed. We were able to show that this specific population of newly recruited macrophages is essential for recovery following spinal cord injury.

Now we were able to demonstrate that, contrary to prevailing medical wisdom, the newly recruited cells are not just additional troops, swelling the ranks of the native immune cells in the spine, the resident microglia. Contrary to the view of our detractors, these cells were not contributing to the ongoing inflammation within the injured spinal cord, but rather were secreting anti-inflammatory substances that brought the inflammatory process under control. It was these cells that provided the injured cord with what it had been missing: a regulated immune response that occurs in the proper sequence and in synchrony, and dials down promptly.[25] It is these cells that make sure that the physiological inflammation is terminated on time before it turns into pathological inflammation.

We discovered that the same macrophages are also the missing link in regulating the scar, namely ensuring its timely removal. As we have seen, since the 1980s researchers had believed that scarring presented an additional obstacle to spinal cord regeneration and healing. Attempts by several groups to get spinal cords to regenerate by preventing scar formation had met with mixed results. Our own studies, initiated by Asya Rolls (at that time a graduate student in my team, now an independent scientist at the Technion) and Ravid Shechter, reconciled these inconsistent results by demonstrating that the scar tissue in the central nervous system is essential in the first days following the injury. At this stage, it is preventing the spinal cord from regrowth in a chaotic microenvironment. It also modulates the immune response and helps to reorganize the tissue at the damage site. After the first few days, the scar becomes an obstacle to healing and should be removed. We found that macrophages interact with the scar and secrete proteins that break it down when it is no longer needed, thereby allowing regrowth and regeneration.[26]

During all these years we were puzzled why macrophages that are numerous in the blood are not more effectively recruited to the injured spinal cord. Is it because of the blood-brain barrier that blocks their entry? If so, should we break an essential barrier for the benefit of healing the severed cord? We learned that the recruited immune cells don't need to break the blood-brain barrier in order to find their way to the injured spinal cord. Instead, they pass through a special gate at the interface between the central nervous system and the circulation, the brain-immune communication border that we introduced in the previous chapters, the brain's choroid plexus. We learned that this unique blood-central nervous system interface is not an inert structure but can act as a gate that selects and filters the cells that enter the central nervous system. This gate even has the ability to control the properties of the incoming cells, affecting the way they function within the nerve tissue. The finding that immune cells pass through this gate on their way to the injured spinal cord surprised us, as this gate is located in the brain, far from the injury site at the spinal cord. In fact, this remote entry, rather than entry at the lesion site that occurs in all other non–central nervous system tissues, might explain why recruitment of immune cells to the central nervous system is so limited. We learned that entry through this route, however, although limiting the number of recruited immune cells, provides a "safer" mechanism, disengaging the fragile nerve tissue from the circulation. Does this understanding imply that animals with longer necks, such as giraffes, have more difficulties in regenerating their cords? We will probably never know![27]

Equipped with our new data, ImmunoBrain Therapies, a

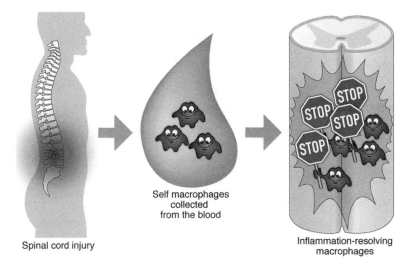

Figure 14. Immune cell–based therapy for spinal cord injury (ProCord).

start-up company in which Eti Yoles, a former postdoctoral fellow in my laboratory, heads research and development, has decided to go back to the clinic with an improved macrophage therapy. The human cell population equivalent to the essential mouse macrophages has been identified, and is now being prepared for the next round of human trials.[28] (Figure 14)

While researchers are still searching for the "magic bullet" that will promote regrowth of severed axons, it is clear that implantation of macrophages can be used to treat multiple aspects of spinal cord injury. It helps control local inflammation, removes the scar tissue, and supplies molecules that support regrowth, altogether reducing secondary degradation following injury and creating conditions favorable for regeneration and repair.

In August 2010 I gave a lecture at the University of Canberra, Australia. In his introduction to my lecture Julio Licinio, now deputy director for translational medicine and head of the Mind and Body theme at South Australian Health and Medical Research Institute (SAHMRI) in Adelaide, South Australia, commented on the findings of our group: "I was absolutely [perplexed]; I had never seen something like that in science, and I literally had tears rolling down my face. . . . [Schwartz] makes the blind see and the paralyzed walk."[29]

The challenge is, of course, making this a reality for humans. We are determined to get there, but there is much work ahead. We are always keenly aware that promising breakthroughs in animal models rarely make the transfer to human patients—and they are many. But as our studies progress, our clinical trials show promising results, and the global scientific community increasingly accepts our approach, we believe we are on the right track.

5

A Vaccination to Prevent Blindness

C an you imagine how your life would change if suddenly you were unable to see? A traveling exhibition called Dialogue in the Dark might give you a clue. You enter a room and are given a cane and a short preview about blindness and basic rules of how to act in the dark. Then the lights go out and you enter an unfamiliar world. You can see absolutely nothing. In this tour in the dark, you confront simple daily tasks that become much less simple when done in complete darkness. You have to walk down stairs, cross what sounds like a heavily trafficked road, buy fresh goods in the market, order coffee (and pay for it), and find your way in the subway, among many other everyday experiences. You learn to use your other senses: You exercise caution upon hearing the sound of an approaching train or a car, you find your way to the coffee shop by the strong scent of coffee, and you choose your vegetables carefully by smelling, tasting, and touching. But most of all you realize that you have to trust the people around you. What makes this experience even more unusual is that the ones who guide you through it feel very much at home in this world—they are blind. For approximately one hour a reversal of roles takes place: Now you are dependent on your blind guide, who helps you find your way in the dark and keeps you safe. Soon you learn that blind people are no more disabled in the light than you are in the darkness.[1]

Think about the location of your eyes in your head. Your eyebrows shield your eyes from rain, sweat, direct injury, and sunlight. Your eyelashes screen out dust, pollen, and other particles, while your eyelids and tear ducts keep the eyes moist. All of these structures have been retained throughout mammalian and primate evolution, indicating how precious eyesight is, and why, for most animals, sight is essential for survival. After all, in the animal kingdom, the ability to see is one of the traits shaped by natural selection. For blind primates, finding food or shelter, as well as fighting off or escaping a predator, becomes virtually impossible, and visual communication is important within and among species, used in everything from bluffing a rival to wooing a mate. An animal that depends on sight and is suddenly blinded is unlikely to pass along its genes.

In people, vision may be impaired by many different conditions, and vision loss is a priority on the public health agenda. According to World Health Organization estimates, as of 2013 there were 285 million individuals worldwide with some form of visual impairment, 39 million of them blind. More than 90 percent of the world's visually impaired people live in developing countries, with four out of five people losing their sight unnecessarily due to conditions that are either preventable or treatable. The avoidable blindness in these countries is related to poverty, malnutrition, and lack of education and appropriate health care services. Various nongovernmental organizations are trying to reduce these numbers by providing eye-care services in developing countries, sharing expertise and data, and encouraging research. But vision may be inevitably lost due to degenerative diseases, congenital or age related. Can the immune system protect us from such diseases that lead to blindness?[2]

A UNIQUE ORGAN CONNECTED TO YOUR BRAIN

When light enters the eye, it passes through the cornea, the transparent front part of the eye. It then passes through the iris and the pupil, which regulate the amount of light allowed to enter the internal structures of the eye. Muscles around the eye adjust the lens, bringing the object into focus on the retina. The retina has the shape of a bowl, and in humans it is about one one-hundredth of an inch thick. It is composed of layers of nerve cells. The nerve cells of the outermost layer of the retina capture light. The innermost layer of the retina contains cell bodies of the optic nerve, the retinal ganglion cells, whose fibers are collected in a bundle. This bundle of nerve fibers leaves the eye as the optic nerve. These optic nerve fibers deliver visual information from your retina to areas in your brain that process visual information, where the image that you see is interpreted. You recognize a friend's face, a stop sign, the change a cashier is handing you.[3]

Our understanding of how visual perception is established in the brain was dramatically advanced by the Nobel Prize–winning work of a Canadian, David Hubel, and a Swede, Torsten Wiesel. Their research began in the mid-1950s and lasted a quarter of a century. In their experiments, Hubel and Wiesel found that neurons in an area of the brain called the visual cortex respond to various stimuli delivered from the eyes, cooperating to create a complex representation of the world from a steady stream of visual information.

In one of their classic experiments in the early 1960s, using newborn kittens, Hubel and Wiesel sewed shut one eye of each kitten and left that eye closed for the first three months of life. Then the stitches were removed. All the kittens became blind in the eye that had been sewn closed, though the experimenters could observe no changes in the eye itself. The only change they found was in the kittens' brains. In the visual cortex, they found that nerves devoted to receiving input from the normal eye had invaded the areas typically assigned to the other eye. Sewing shut one eye had rewired the kittens' brains. This pioneering work demonstrated that processes occurring in the eye shape the growing brain, emphasizing the strong connection between these two organs.[4]

Indeed, the retina forms a continuous part of the brain, segregated from it during early embryonic development but maintaining its connections with the brain through the optic nerve. Although located on the farthest reaches of the central nervous system, retinal ganglion cells display typical properties associated with nerve cells in the brain. They share the same basic structure: cell body, dendrites, and axons (see the Primer for more details). These axons are also covered by myelin, an insulation material that covers all axons, and that allows signals from the eye to the brain to be conducted efficiently. So we see that the cell bodies of the optic nerve, the retinal ganglion cells, closely resemble neurons found in other parts of the central nervous system such as the brain and spinal cord. This notion has inspired research in two distinct but complementary directions: studies of the eye in order to understand some processes that occur in the brain, and gathering knowledge in brain research to study the eye, as we will show below.[5]

AN EYE APART?

Although anatomists consider the eye as an extension of the brain, modern medicine in the West has dealt with the organ as if it were a separate entity. Encased as it is by blood-tissue barriers and shielded by various mechanisms that restrict immune responses, the eye was believed to be hardly influenced by the health of the rest of the body and cut off from the immune system. Medical researchers and clinicians alike, in treating diseases of the eye as if it were an autonomous organ, have ignored two possibilities: first, that eye diseases might share many mechanisms with classical neurological ailments or may even be an early symptom of such brain ailments; and second, that the eye may be influenced by immune molecules circulating in the bloodstream.[6]

Over the past two decades my group has pioneered a new direction in research into eye pathologies. First, we proposed that the progression of chronic eye diseases resembles the progression of brain diseases, regardless of the primary cause. Later, we showed that degenerative diseases of the eye, such as glaucoma, might be treated by recruiting assistance from the immune system.

A DOMINO EFFECT

During the 1990s, neuroscientists studying central nervous system trauma proposed that under certain conditions when the brain is damaged, a process of secondary degeneration takes place, as we saw with the spinal cord injury. Whatever the initial cause, degeneration of nerve cells could become self-perpetuating, in a sort of "domino effect," eventually affecting adjacent neurons that had been spared in the initial injury. The culprits were the same factors we had found to be toxic to neurons in the brain and spinal cord: such cell-threatening components as free radicals, unbalanced levels of neurotransmitters, and abnormal accumulation of "junk": broken cell parts and cell waste. This toxic junkyard also *lacked* certain neurotrophic factors, such as brain-derived neurotrophic factor (BDNF) and nerve growth factor (NGF)—factors that support cell growth, prolong cell survival, and promote regeneration of axons. Under these toxic conditions, nerve cells appear unable to repair themselves.[7]

In the late 1990s we were looking for ways to study this domino effect of secondary degeneration in the central nervous system. Having recognized the relatively accessible nerves located in the eye, we decided to create a rodent research setup of eye injury that would allow us to study the death process of such neurons. We hoped we could learn from this setup how neurons respond to injury in general, in other parts of the central nervous system. Together with Eti Yoles, then a young postdoctoral fellow, we adopted the injured optic nerve in a rat as a research setup to evaluate the domino effect; we wanted to determine whether neurons that were spared in the injury but died later did so because of the toxic microenvironment created by the injury, and whether this delayed death could be prevented by a therapeutic strategy. To quantify cells that survived, we injected a dye behind the lesion site in the optic nerve, close to the brain. Only intact axons could take up the dye and transport it through the lesion site, all the way to the optic nerve cell bodies located in the retina. When we looked at these retinas under the microscope, we could count the colored cells, which represented only surviving neurons.

This technique gave us a relatively easy way to measure how many neurons were dying from the initial injury to the optic nerve versus how many were dying as a result of the domino effect. We were astonished to discover a self-perpetuating process of degeneration. Similar to the domino effect associated with brain damage, degeneration in the optic nerve expanded to include nearby fibers that had escaped the original lesion.[8]

Excited as we were with these findings, eye diseases were certainly not our expertise, and so we turned to Michael Belkin, an expert researcher in ophthalmology at the Sheba Medical Center at Tel Aviv University. Based on our discussions with Belkin, Eti Yoles and I realized the magnitude of our findings with respect to diseases like glaucoma, one of the leading disorders responsible for vision loss and blindness.

"Glaucoma" refers to a group of related diseases, all of which involve damage to the optic nerve and lead to cell death in the retina. For decades, glaucoma was perceived as an eye disease associated solely with damage created by elevated pressure inside the eye, so-called intraocular pressure. Intraocular pressure depends on the clear fluid residing inside the eyeball that bathes and nourishes the eye tissue. This clear fluid is produced by the ciliary body, a structure located behind the iris. It then flows through the pupil and fills the front space of the eye. The fluid drains from the eye to the blood through a structure called the trabecular meshwork, located at the intersection between the cornea and the iris. If too much of this fluid is secreted into the eyeball or if it drains too slowly, pressure begins to rise. Excess pressure can damage the optic nerve, leading to glaucoma.[9]

Before our observation that a single injury to the optic nerve can lead to a cascade of degradation, my team had never worked on glaucoma; following this observation we became "glaucoma experts." We learned that the therapies used to treat glaucoma, at that time and even now, are based on drugs that reduce pressure inside the eye. Yet over the years, physicians often have observed that even after the pressure within the eyeball is reduced to normal or even below-normal values, glaucoma can still progress, with the patient continuing to lose vision, sometimes to the point of blindness. This observation left the researchers and the clinicians with a puzzle. What were the missing links, what are the additional, unknown factors besides pressure that might be at work in glaucoma—and perhaps other diseases that involve death of nerve cells?[10]

Our studies allowed us not only to establish the eye as a research system for monitoring loss of neurons that are not directly damaged but are rather affected by the domino effect, but also to propose for the first time that one of the underlying problems in glaucoma is the self-perpetuating degeneration of retinal cells.

My group found that, beyond the negative effect of intraocular pressure, disease aggravation in glaucoma is caused by the degeneration-causing chain reaction similar to that seen in other neurodegenerative disorders, as well as in injuries to the central nervous system. Even after the primary cause (the increased pressure inside the eye) is alleviated, the process of degeneration continues, resulting in irreversible damage to the eye. We were intrigued to learn that substances suspected of causing a domino effect are found in the eyes of patients affected with glaucoma. Glutamate is one of them. Under normal conditions, glutamate is an essential brain-derived communicating molecule. Yet patients with glaucoma have higher than normal levels of glutamate in their eyes. Our conclusion was that reducing intraocular pressure might not be sufficient to arrest the disease. In 1996 we were the first to propose that glaucoma, a chronic disease of the visual system, might benefit from neuroprotective therapy—treatments that would rescue neurons—in addition to the pressure-reducing therapy.[11]

Our proposal was so influential on the community of ophthalmic research that since we first came up with our idea, in 1996, neuroprotection has become a prominent direction in the search for a cure for glaucoma. Soon, everyone was looking for neuroprotective agents.

At conferences and international forums devoted to the research and treatment of glaucoma, researchers continued to debate the identity of such agents. One of the drawbacks in translating such a therapeutic strategy into the clinic is that, unlike pressure reduction therapy, in which the only desired outcome has been lowering the intraocular pressure, neuroprotection requires overcoming numerous factors associated with the domino effect. Another limitation in the translation of neuroprotective strategies is the need for a long-term follow-up to detect any effect, due to the slow progression of the disease together with the difficulty of detecting any effect of neuroprotection.[12] We suspected that the solution might be one that will augment a physiological protection process that addresses multiple domino-causing factors. Is the immune system up to it?

———∿∿∿∿∿∿∿———

When I realized that immune cells are actually needed for protection and repair of the central nervous system, I radically rethought the role of the immune system in the domino effect. Contrary to the common view that the immune cells are only a threat to the central nervous system, which should be suppressed under all pathologies, we began to see them as cells that should be harnessed if we were to have any hope of halting degeneration. This was a key discovery that revealed the essential contribution of immune cells within the central nervous system. We found that the immune system helps to cope with damaged nerves. This understanding, applicable also to ravages in the brain and spinal cord, actually originated from our study of the optic nerve and retina.

For example, we found that rats suffering from immune deficiency had fewer surviving retinal neurons after optic nerve injury compared with rats with healthy immune systems. These immune-deficient rats seemed to cope poorly with the domino effect, with reduced ability to rescue spared nerve cells. Those nerve cells that were damaged by the mechanical injury inevitably died, but those that were spared by the primary injury could not survive the process of self-perpetuating degeneration when the immune system in the circulation was weakened. Further studies revealed that a unique population of immune cells is responsible for the rescue of the nerve cells. Such cells can recognize the body's own proteins present at the site of damage in the central nervous system. Effectively, these were the protective autoimmune T cells, as we have described.

When we injected such protective autoimmune T cells into animals whose optic nerves had been crushed, more neurons within the retina (whose axons make up the optic nerve) survived than in animals that were not given the injection.

This fundamental discovery led us to propose that, as opposed to autoimmune disease, in which the autoimmune response is extremely vigorous, continuous, and destructive, a moderate and regulated autoimmune response that stops when it is no longer needed protects cells within the optic nerve, as well as in other tissues of the central nervous system. In 1999 we named this revolutionary theory Protective Autoimmunity.

Through this discovery and the emerging concept, our team changed the common perception of autoimmunity, from a phenomenon that should be aggressively suppressed by all that medicine has to offer to one that is crucial to health when it is well controlled. This concept was a major milestone in our research. It set the foundation for numerous additional studies that demonstrated that autoimmunity can be beneficial, and more specifically that this response is needed to facilitate repair and tissue maintenance in other parts of the central nervous system. In fact, it led to the discovery of the "immune cells of wisdom."[13]

A DELICATE BALANCE

Elated as I was by this insight, we still had a lot of questions. If autoimmunity is indeed a natural response that has been evolutionarily adopted to maintain human health, why do the optic nerve and other tissues of the central nervous system cope so poorly with injury and disease? Why don't they recover spontaneously? Are we limited in our ability to mount a protective autoimmune response in the central nervous system? The answer lies in the *regulation* of the immune response. For such an immune response to succeed in protecting neurons, the activity and quantity of immune cells, the site of their deployment, and the timing and duration must be tightly regulated. As we have mentioned, the central nervous system has evolved various mechanisms that restrain immune responses to prevent runaway inflammation. However, such restriction also limits the recruitment of beneficial immune cells, curbing the assistance that is so crucial following damage to the central nervous system.

Specifically, the natural autoimmune response is under tight control to allow the benefit without the risk of imposing autoimmune disease. Based on our understanding that autoimmunity is a purposeful protective response, such a tight control seems a drawback under pathological conditions, when an autoimmune response is needed. According to our view, it is only when the immune response escalates out of control that it results in autoimmune disease. The same immune cells that promote healing, when they are not properly regulated, instead cause cell death.

Understanding that the natural levels of autoimmune cells are restricted, and that under disease conditions there might be a need for more of these cells, we searched for a way to safely augment their levels while avoiding any risk of causing symptoms of autoimmune disease. Our group designed a unique vaccination therapy. As in the case of the vaccine for the mind, this vaccination also consisted of a synthetic peptide that mimics a fragment of a protein that is common at the site of damage, such as myelin. This compound can induce the activity of weak autoimmune cells; just enough to allow them to protect neurons but not enough to evoke an autoimmune disease.

We took rats with damaged optic nerves and vaccinated them with these kinds of central nervous system—mimic compounds. This treatment proved effective in halting the domino effect of cell death in the optic nerves of the rats. We also showed that the vaccination was effective in animals in which glaucoma-like disease was induced. These vaccinations were a major breakthrough, as they allowed us to engage the benefits of the immune response in eye diseases.[14]

HOW THE NEUROPROTECTIVE VACCINES WORK

Exactly how do these vaccinations protect the nerves? We found that the autoimmune T cells fight off the neurotoxic factors present after injury by creating a supportive environment, enriched with protein factors (such as BDNF and NGF) that support cell growth and survival. As we saw in animal models of spinal cord and retinal injury, these immune cells also orchestrate recruitment of macrophages, the key players that help shut down the domino effect by stopping local inflammation, protecting neurons, and helping to renew cells.[15] Again, a complex immune cell circuit is orchestrating the proper and safe response to preserve sight.

We still don't have all the answers. We aim to decipher exactly how the vaccinations work. Regardless of the mechanism of action, these vaccinations seem to promote a balanced immune response that stops the domino effect and reduces further death of neurons. The response is strong enough to promote healing, and, of equal importance, is terminated in time to avoid a risk of evoking autoimmune disease as a side effect (figure 15).

Studies have shown that many eye diseases in addition to glaucoma can be relieved by boosting the body's immune response. Regardless of the primary cause of the disease, whether it is due to an inherited mutation, as in the case of retinitis pigmentosa; accumulation of cellular debris or uncontrolled growth of blood vessels, as in age-related macular degeneration (AMD); or an uncontrolled autoimmune response, as in posterior uveitis (inflammation at the back part of the eyeball), the immune system is always a prime actor. In some cases the immune response is insufficient to halt the

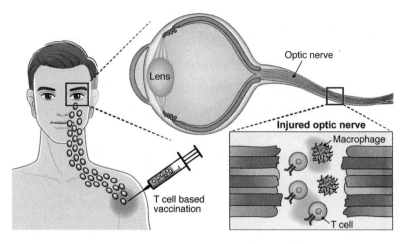

Figure 15. Vaccine for preserving sight: Therapeutic vaccination based on protein fragments which mimic self-compounds leads to recruitment of protective immune cells to the injured site. These cells protect the eye.

disease, and in other situations, the response is not properly controlled and becomes destructive. Immune intervention via vaccine holds great promise. By restoring the proper immune balance and maintaining supportive immune activity, such therapies offer new hope for preventing blindness.

In 2002, I received the Friedenwald Award in recognition of my pioneering work in ophthalmology. The Friedenwald Award was established in 1957 as a memorial to Jonas S. Friedenwald, a distinguished researcher whose contributions encompassed the entire field of ophthalmic investigations. His pioneering studies on the pathogenesis of glaucoma, corneal wound healing, and diseases of the retina laid the groundwork for future generations of investigators. I was honored, but knew that the way from the laboratory to the clinic is still long.[16]

In searching for an appropriate compound to incorporate in a safe therapeutic vaccine for humans, we chose Glatiramer acetate, a synthetic chain of amino acids that is slightly similar to brain myelin proteins. We thought that this compound would allow us to safely activate the immune system for neuroprotection, much the way we use a dead or weakened virus to vaccinate against a live viral pathogen. Given in a daily regimen, Glatiramer acetate, better known as Copaxone, is an FDA-approved drug for the daily treatment of multiple sclerosis (MS). We adopted this compound and decided to develop it as a vaccine for glaucoma, which required further research to determine a proper regimen of administration. After establishing that such a therapeutic vaccine is an effective neuroprotective treatment in animals, provided that it is given weekly or monthly, in contrast to the daily regimen given to MS patients, we went on to develop this treatment for humans. The license to develop this compound as a treatment for glaucoma patients and, as you will see in the next chapter, also for other chronic neurodegenerative diseases, was given to Proneuron. Proneuron agreed to collaborate with Teva Pharmaceuticals, which had developed and marketed this compound as Copaxone for treating MS. Unfortunately, due to erroneous commercial considerations, a financial conflict of interest, and a long legal dispute, the treatment has yet to be developed for glaucoma patients.[17]

Despite this setback, our efforts to unravel the potential benefit of the immune system and the great human and financial capital invested in designing the vaccination have proved worthwhile. A clinical trial performed in Brazil in 2006–7 showed the success of such a vaccination in preserving retinal neurons in thirteen patients with diabetic retinopathy. This complication of diabetes damages blood vessels in the eye, destroys the retina, and can lead to blindness. Patients in the study group received four weekly injections of the vaccine, and their retinal nerve fibers were found to be significantly thicker than those of patients in the control group, who received a placebo. That thicker layer of retinal nerve fibers indicates a higher survival rate of retinal neurons, and potentially a greater preservation of sight. As opposed to most vaccines that act to prevent illness, this vaccination served as therapy. It did not eradicate a foreign pathogen but rather dealt with internal enemies by boosting the immune system in order to arrest degeneration and preserve retinal nerve fibers.[18]

You are probably familiar with the saying "the eyes are the window to the soul." Our group views the retina as the window to the central nervous system, a useful tool for investigating the brain and diagnosing its diseases. Many of the pathological features typical of eye diseases are shared by several other disorders of the central nervous system. In addition, several well-defined neurodegenerative conditions that affect the brain and spinal cord, such as stroke, multiple sclerosis, and Alzheimer's and Parkinson's diseases, can show up in the eye, and these eye symptoms often emerge before conventional diagnosis of such disorders. Moreover, being located outside the skull, the eye is relatively accessible for manipulation and imaging in healthy study subjects as well as in patients, making it a viable model for the study of processes within the central nervous system in both health and disease.[19]

The domino effect of degeneration is shared among all degenerative diseases of the central nervous system. Given the many similarities among the various parts within this system (eye, brain, spinal cord), we suspected that a treatment that could rescue retinal neurons might also save brain or spinal cord nerve cells. With this goal in mind, we began extrapolating our findings from the visual system in an attempt to treat injuries and disorders in the brain and spinal cord. With minor modification, depending on the organ in need, we were able to develop potential therapies for conditions such as Alzheimer's disease and ALS, as we shall describe in the next chapters.

Over the years, we learned that we could augment levels of autoimmunity not only by vaccination but also by reducing the cells that keep autoimmune cells under control, the "brake" cells. Thus we have alternatives in immune-based therapies: We can either push down the gas pedal or release the brakes, ultimately yielding similar results. The choice, however, requires an understanding of where the problem lies in the immune system in each disease. This is where we are now, developing immune-based therapies for glaucoma and age-related macular degeneration, designing the best immune strategy, with the common goal of customizing the immune response for different eye ravages.

6

Alzheimer's Disease and Lou Gehrig's Disease (ALS)

W e have learned that the immune system can help heal a sudden injury to our spine, mend a crushed optic nerve, or repair damage to our retinas. How does it deal with chronic brain disorders? Could it help the brain cope with the continuous burden of neurodegenerative disease? We know now that in many diseases, including cancer, there is a component of inflammation. Now that we understood that uncontrolled inflammation takes place in neurodegenerative diseases, we asked what the way our immune system deals with cancer could teach us about neurodegenerative diseases such as Alzheimer's disease and the fatal motor neuron disease amyotrophic lateral sclerosis (ALS)?

It may seem nonsensical to compare diseases in which cells die, such as neurodegenerative diseases, and those in which cells multiply out of control—for example, cancer. It may sound weird and even incomprehensible. But that is not the case. For decades cancer has been perceived as the most threatening disease of the millennium. Almost all of us have encountered this disease some way or another: by caring for a sick parent or partner, or supporting a friend who has been diagnosed. The American Cancer Society warns against a rising global cancer epidemic, with reports on nearly eight million cancer-related deaths each year worldwide, projected to exceed thirteen million by 2030. Major efforts are under way to find ways to diagnose cancer earlier, which critically affects prognosis, and to come up with new treatments. Many of these efforts are invested in trying to understand the primary cause of various cancers and the way they progress into full-blown disease.[1]

Over the years, it has become clear that cancers do not depend solely on cancerous cells. Their emergence, as well as their progression, depends on the strength of the immune system. The immune system is on constant lookout for aberrant cells and nascent tumors, trying (mostly successfully) to eliminate these abnormalities before they lead to disease. The prevailing dogma has been that inflammatory diseases and cancer are completely unrelated, if not mutually exclusive. Now it is clear that local, chronic inflammation is the basis of many tumors. It is the result of the struggling immune system trying to eradicate the tumor and prevent disease progression. The tumor, on the other hand, deceives its attacker by suppressing the immune system. In this way, rather than being eradicated by immune cells, it evades the immune response and continues to grow uninterrupted. When the immune system is not strong enough, the tumor gains the upper hand. It can seize the opportunity, emerge, and progress.[2]

Here we will show you that neurodegenerative diseases such as Alzheimer's and ALS share similarities with tumors. Like tumors, they may stay dormant long before their onset due to continuous patrol of the immune system. We are proposing that when such immune patrolling is out of balance, neurodegenerative diseases emerge. By no means am I proposing that the immune system is necessarily the primary cause of such diseases, but rather that the failure of the immune system allows the emergence and fast progression of these diseases. In this chapter we shall introduce you to the cross-talk between such neurodegenerative diseases and the immune system. What happens when the immune system can no longer contain these diseases, and how can we harness immune cells to prevent, delay, or assist in resolving the death of neurons?[3]

———~~~vvvvvvvv~~———

A 2011 international survey, including the United States and four European countries, revealed that Alzheimer's is one of the biggest health-related fears, second only to cancer. In this survey, conducted by Alzheimer Europe and the Harvard School of Public Health, the fear of getting Alzheimer's was ranked higher than that of such life-threatening diseases as heart disease, stroke, and diabetes.[4]

Considering the devastating outcomes of this disease, this is not surprising. Alzheimer's affects the brain, leading to brain dysfunction, culminating in the death of nerve cells in brain regions controlling memory, thought, and language. The neuronal dysfunction seriously harms a person's ability to think clearly, remember things, and carry out daily tasks. Over time, as illness gets worse, a person with Alzheimer's may fail to recognize family and close friends, become confused and wander from home, experience personality changes, even become aggressive. Patients may experience difficulties communicating as it becomes difficult to speak, read, and write. In the end, the patient may need round-the-clock care.[5]

The progress of the disease is agonizing both for the patient's family, who watch their loved one deteriorating, and to the patients themselves, who gradually lose their clarity of thought. Such pain can be best expressed by a quotation from the diary of an Alzheimer's patient: "Every few months I sense that another piece of me is missing. My life . . . my self . . . are falling apart. I can only think half-thoughts now. Someday I may wake up and not think at all."[6] Now imagine the opposite extreme: a disease that keeps your mind perfectly operative but gradually destroys your physical body, leaving you imprisoned. This is the case in amyotrophic lateral sclerosis (ALS), also known as Lou Gehrig's disease, after the famous New York Yankees first baseman who died from this disease in 1941 at the age of thirty-seven. In this condition, the motor neurons, neurons that deliver information from the brain and spinal cord to the muscles, progressively die, resulting in a gradual weakness of the muscles, which eventually affects chewing, swallowing, speaking, and breathing and finally culminates in complete paralysis and death, usually within five years of diagnosis.

ALS patients clearly notice their own physical deterioration and are completely aware of the way their lives are going to end. As the disease develops, the person needs increasing help with daily routine tasks: getting out of bed, eating, walking, taking a shower. The disease imposes an enormous emotional challenge on both the patient and his family, who must deal with disability and anticipate further deterioration in the future, culminating in an unavoidable death sentence.[7]

Although Alzheimer's and ALS seem diametrically different in their nature, they have something in common. Though some of their symptoms can be managed with drugs, both are incurable.

Despite enormous research efforts, we don't yet know what causes Alzheimer's disease. There are a few theories, including the reduction in certain brain-derived communicating molecules, such as acetylcholine, that are used by the brain in areas devoted to memory and learning, like the hippocampus; so-called senile plaques, the accumulation of proteins around neurons; the deposits of certain proteins within the bodies of neurons; and neuron inflammation that involves toxic, inflammatory compounds that damage neurons.[8]

Several drugs have been approved for the treatment of Alzheimer's, but their efficacy is limited. Most of these drugs focus on relieving some symptoms and improving the patient's quality of life. None can halt or reverse the progression of this disease. Moreover, the drugs are effective only for a short term, after which they cease to work.[9]

ALS presents an even more dire prospect. Although the medical community has recognized the disease for more than a century, not until the early 1990s did scientists begin to decipher some of its potential causes. A hereditary factor exists in some ALS patients; several genetic mutations were found in these families. One of the most common mutations is the one that damages superoxide dismutase 1 (SOD1), a protein that protects cells from free radicals. Mice genetically engineered to carry the SOD1 mutation display degeneration of their motor neurons and are the most studied animal model of ALS. More recently, an additional mutation associated with ALS was identified in the C9orf72 gene. This gene includes a short DNA sequence that is repeated from two to twenty-three times in healthy individuals but between seven hundred and sixteen hundred times in some ALS patients. The function of the protein encoded by this gene is currently unknown, but researchers suggest that the expanded sequence repetition interferes with the process of converting the genetic code into a functional protein, resulting in aggregation of genetic material and exacerbated toxicity in neurons. To date, this is the most common genetic abnormality found in both familial and nonhereditary ALS cases. Unfortunately, even though several genes have been identified, therapy is still proving elusive.[10]

Moreover, while genetics may predispose people to ALS, there are undoubtedly environmental factors that determine whether, when, and how the ALS genes are switched on. Those factors remain unidentified. For patients who do not have a family history of the disease, theories range from an excess of brain-derived communicating molecules—neurotransmitters—which switch on neurons, to dysfunction in cell metabolism, to problems with nerve conduction, and to immune cells that don't function normally. There is even a theory that irregular protein structures, known as prions, may be to blame, as is the case in Creutzfeldt-Jakob disease, the human version of spongiform encephalopathy or "mad cow disease."[11]

Because theories for ALS development represent such a wide spectrum of potential causes, scientists are looking for aspects of lifestyle shared among ALS patients that might shed light on the way this illness develops. Excessive exposure to certain metals or chemicals in the workplace, repeated viral infections, smoking, and even strenuous physical activity and military service are all associated with increased risk of ALS.[12] Various studies have confirmed an almost doubled risk of ALS among military personnel who had been deployed to various conflict areas across the world, leading the U.S. Department of Veterans Affairs to declare ALS a disease connected to military service. Researchers have also noted that ALS might be overrepresented among athletes, with Lou Gehrig being the most famous example.[13]

But no one yet knows why these groups are at higher risk of getting ALS. With no reliable information about what might cause the disease, therapy seems out of reach for the near future.

The only drug approved for the treatment of ALS by the United States Food and Drug Administration (U.S. FDA) is 6-(trifluoromethoxy) benzothiazol-2-amine, trade name Rilu-zole. This drug helps patients cope with the excess of excitatory neurotransmitters. Unfortunately, although it alleviates symptoms for selected patients, and perhaps delays reliance on a ventilator, the drug has only a modest effect on prolonging patients' lives, typically giving them another two or three months.[14]

Over the past two decades, neurodegenerative diseases have often been linked to chronic inflammatory processes within the brain or the spinal cord tissue. Again, as we have seen in this book, it was commonly believed that in the diseased central nervous system, overactivated immune cells could create a vicious self-perpetuating cycle resulting in a prolonged, unregulated inflammation that drives the chronic progression of such diseases.[15]

Based on this assumption, many clinicians treated Alzheimer's, ALS, and other neurodegenerative diseases with anti-inflammatory drugs and steroids, yielding conflicting results.

The immunosuppressant Minocycline showed promising results in mice with ALS but showed no benefit in human patients, and in some cases even exacerbated the disease. A similar hope that anti-inflammatory drugs might prevent the onset of Alzheimer's didn't pan out; the drugs failed to help patients.[16]

According to my theory, at the root of this approach lies a misconception about the nature of inflammation. As we have emphasized, every condition associated with inflammation in the central nervous system has traditionally been linked to autoimmune inflammatory pathologies, as in multiple sclerosis, in which the circulating immune system attacks the brain.

Now we know that inflammation within the brain starts out as a good thing: It is the body's physiological response, a way of using the immune system against risk factors within the brain that may cause chronic neurodegenerative conditions. When this immune response isn't switched off on time, it becomes an unresolved process that can contribute to disease progression. With more and more pro-inflammatory effector immune cells engaged in a desperate and unsuccessful attempt to contain the disease, it seems like a vicious cycle that further exacerbates the disease. While it begins as a defense process, physiological inflammation can soon become pathological inflammation. The belief of clinical researchers that we can break such a vicious cycle by suppressing the immune system outside the brain has turned out to be an illusion. As we understand from our studies, suppressing the immune system outside the brain to shut off chronic inflammation within brain tissue can be counterproductive, as it denies the diseased brain the immune assistance it needs from the blood. This assistance includes that provided by blood-borne immune cells, recruited to the damaged site, which eventually help extinguish the vicious cycle of inflammation within the brain. Thus general immune suppression might dampen the response of such immune cells that serve as "cease-firing," immune-resolving cells. As an alternative to the ineffective general immune suppression, we suggest augmenting recruitment of the appropriate immune assistance from the blood to the brain.[17]

———~~~~~~~———

What if these very different neurodegenerative diseases all share the same basic pathology: a failure of the immune system outside the brain to send its immune-resolving troops to the diseased brain—the very cells that can bring the local inflammatory fight into an end, and provide supportive conditions for the tissue healing? Starting with that hypothesis, we wondered whether we could heal these diseases by reinforcing this aspect of the immune system.

The so-called senile plaques found in the brains of Alzheimer's patients consist of a toxic deposit of accumulated proteins, both surrounding and within neurons. These plaques are usually associated with severe inflammation.[18]

However, studies have shown that not all immune cells within the senile plaques contribute to disease progression. The plaques consist of a mixture of immune cells, including ones that normally reside in the brain (the microglia) and the newly recruited blood macrophages. The microglia that have failed in plaque removal were found to cause a vicious cycle of inflammation, whereas the blood macrophages removed the plaques. These cells swallow up the proteins forming toxic senile plaques and slow the progression of Alzheimer's disease.[19]

Our own studies have shown that the newly recruited blood macrophages have multiple functions. They produce growth and survival supporting factors, and can contain the local inflammation. These cells make up the "fire brigade" that the diseased brain needs to extinguish the inflammatory "wildfire." In the absence of these macrophages, senile plaques are much more spread out and aggressive in the brain. We can harness the immune cells that shut down the local pathological inflammation and remove the senile plaques. As we have seen in cases of acute central nervous system injuries, newly recruited macrophages from the blood can do just that, providing anti-inflammatory factors and getting the immune response back under control. These studies, although received with skepticism by some of my colleagues, even eight years after we first proposed that blood macrophages are needed for central nervous system repair, have nonetheless paved the way for numerous additional studies that reproduced our results, and researchers are now seeking ways to bring macrophages into the diseased brain. These findings might also explain the failure of anti-inflammatory drugs in treating Alzheimer's disease —they indiscriminately impede all immune cells, including the protective ones, such as the beneficial macrophages.[20]

139

We then asked what it is that keeps the Alzheimer's brain from recruiting those essential macrophages in the first place. Why do we need to intervene? Studies we have completed in the past two years imply that the bottleneck may reside in the site of entry of these immune cells to the brain. In the animal model of spinal cord injury, we found that newly recruited "healing" blood macrophages enter the damaged central nervous system through the brain's choroid plexus, the special communication border between the brain and the immune system that we identified as a selective gate. As we have seen, this site acts by selecting the immune cells entering the brain and by adjusting their function according to the brain's needs. However, we found that under Alzheimer's disease conditions, this gate loses some of its properties, leading to a failure to recruit appropriate immune cells to the central nervous system.[21]

In the case of Alzheimer's, it seems that a malfunctioning of the gate results in insufficient numbers of macrophages being recruited. We asked why this crucial immune gateway is closed, just when immune cell recruitment is most needed. Has the disease process in the brain robbed it of the ability to summon help from the immune system? Or does it reflect a problem with the immune system itself? Whatever the cause, how can we activate the choroid plexus and enhance the recruitment of these essential immune cells?

Through our experiments, we discovered that in individuals suffering from Alzheimer's disease, this gate is shut. Both in normal aging and in Alzheimer's disease, there is a shift in the immune system that tilts its activity toward cells that suppress the immune system. Although these suppressor cells are normally designed to keep the immune response in continuous check, an excessive concentration of them may completely block the immune response. It is like pressing too hard on the brake when starting down a hill. When the balance shifts toward such suppressor cells, it denies the brain's needs for the immune assistance, both by reducing the numbers of protective immune cells and by preventing the brain's gate from allowing entry of these cells, thereby aggravating the neurodegenerative conditions. A similar phenomenon emerged in cancer research when it was found that tumors manage to survive in the body by attracting immune suppressor cells, which shut down the immune response that attempts to eradicate the tumor. The tumors escape from the immune system by exploiting the body's own immune-suppressing cells.[22]

Having elucidated the reasons for impaired immune assistance in Alzheimer's disease, we were still seeking ways to overcome such immune restriction. Through our studies we discovered that one way to overcome the immune suppression is through vaccination, which boosts the level of protective immune cells that recognize brain compounds. Such vaccination results in a surge of healing macrophages to the site of damage. We found that such a vaccination revived the brain's immune gateway and helped recruit vital macrophages.[23]

From a treatment perspective, we decided again to use Glatiramer acetate, a synthetic chain of amino acids that is slightly similar to brain proteins. We took mice that had been genetically engineered to develop Alzheimer's disease and vaccinated them with Glatiramer acetate (a weekly injection rather than the daily regimen given to multiple sclerosis patients). After treatment, these mice showed cognitive restoration, almost complete disappearance of senile plaques, and increased creation of new neurons, compared with untreated mice with Alzheimer's disease. We were amazed to discover that a vaccine could reverse an advanced case of Alzheimer's disease. In contrast to most vaccines, which are usually used to prevent disease—but like the vaccine, described in the previous chapter, used to reinforce retinal nerve fibers—in this case we designed a vaccine as a treatment, which potentially could be given to patients that have already been diagnosed.[24]

We found that the therapeutic vaccination changed the balance of the immune system in the blood, reining in the suppressor immune cells and allowing the protective immune cells to do their job. This, in turn, enabled the activation of the gate to the brain that allowed the trafficking of the essential macrophages and enhanced production of growth factors and anti-inflammatory compounds. Thus a vaccination therapy that can safely boost the protective immune response was found to be successful in the treatment of a neurodegenerative disease like Alzheimer's.[25]

An important discovery was that Glatiramer acetate in a daily regimen, as it would be used for multiple sclerosis patients, was not effective, and was even destructive in the model of Alzheimer's disease. A daily regimen, as opposed to the infrequent regimen we used, increases the suppressor cells, which shut off the autoimmune response. In the Alzheimer's disease brain, immune suppression by these suppressor cells is counterproductive, as it further blocks the brain's gate from recruiting essential "antifire" immune cells to the diseased brain. Since such suppressor cells, however, are beneficial in common forms of multiple sclerosis, our results emphasize that caution should be taken when trying to extrapolate from one disease to another, even if these diseases share some common pathways. Immune suppression is not the ultimate miracle answer to all inflammatory conditions.

———〜〜〜〜〜〜———

Although the causes of amyotrophic lateral sclerosis remain unknown, it has also been linked to an imbalanced immune response. In line with our theory that protective autoimmune cells can alleviate various pathologies of the central nervous system, other researchers have shown that T cells are essential for prolonged survival of ALS mice.[26]

The immune incompetence in ALS could be a cumulative result of several different processes. Our studies have shown that one such process is partly due to excess levels of immunoregulatory cells in the blood (see Primer); we found that ALS patients had increased blood levels of immunoregulatory cells compared with healthy controls. In addition, in ALS patients, there is a general reduction in the rate of immune cell renewal.[27]

Overall, in ALS, as in Alzheimer's disease, it seems that the central nervous system has lost its ability to recruit the appropriate immune assistance. Again, we found that during the progression of ALS the brain's gateway for healing immune cells is hardly activated. There is no increase in the levels of molecules that attract immune cells and assist in their trafficking to the site of damage. This might explain the limited recruitment of immune cells to the central nervous systems of patients with ALS.[28]

In both Alzheimer's and ALS, as well as in aging (see Chapter 2), and probably in many other neurodegenerative diseases, the body fails to cope with the disease because the brain-immune communication is dysregulated just when immune support is most needed. No matter what the source of this immune incompetence, which leads to gate impairment, treatment may come from boosting the recruitment of specific immune cells, according to the needs of the tissue (figure 16).[29]

Overall, our view has challenged the approach of using immunosuppressive treatments for neurodegenerative diseases or countering a single disease factor, such as targeting only the plaques in Alzheimer's disease. Rather, our approach is to strengthen the body's immune mechanisms that globally help by resolving all threats.[30] This can be implemented in the case of Alzheimer's disease by giving the patients Copaxone in an infrequent rather than daily regimen in order to boost immunity. Clinical studies are needed to fine-tune the regimen for patients.

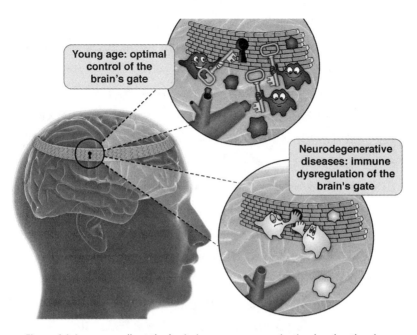

Figure 16. Immune cells at the brain-immune communication border, the choroid plexus, under different conditions. At a young age, sufficient immune cells reside at the brain's border, maintaining daily brain functions. Under neurodegenerative conditions, immune cell dysfunction at this border, which serves also as a gate for the entry of healing immune cells, leads to premature aging of the brain.

But as we have been reminded by the different regimens of Glatiramer acetate indicated in the treatment of relapsing-remitting multiple sclerosis and Alzheimer's disease, each disease is unique, and caution should be taken when using an immune-based therapy that works against one disease to treat a different pathology. We found that vaccination with Glatiramer acetate that was effective in animal models of Alzheimer's disease was not effective in mice engineered to develop ALS-like symptoms. Data from our 2015 studies suggest that in ALS a stronger vaccine regimen is needed. Such a regimen activated the central nervous system entry site, the choroid plexus, enabled trafficking of immune cells to the central nervous system, and lengthened the lives of mice with ALS-like disease.[31]

Designing the appropriate treatment for human patients is essential in order to re-create the success achieved in animal studies, and to avoid adverse events. This is not a trivial task. In 2002 a clinical study of patients with Alzheimer's disease treated with synthetic full-length Amyloid peptide (a prevalent component of senile plaques) was suspended due to a side effect manifested by a devastating immune response reported in the brains of some of the treated patients. Follow-up studies revealed that the treatment had markedly cleared plaque from the brain and reduced functional decline. These results highlight the potential benefit of such an immune-based therapy, provided that the robust detrimental immune response side effect can be avoided. An improved treatment based on the same principles was recently developed and is being evaluated in clinical trials. Keeping safety in mind, our suggestion of using Glatiramer acetate as a potential treatment for Alzheimer's disease is attractive, as this drug is already FDA approved and used daily by multiple sclerosis patients. Nevertheless, the treatment regimen and prescribed dosage should be carefully designed before testing begins in Alzheimer's patients.[32]

Data that we have accumulated in the past few years have suggested an alternative approach, currently under development toward implementing our results to patients. This approach would use compounds that can boost the immune response by controlling the immune "brake" cells, the cells that restrict the immune response.

By identifying the key player behind various neurodegenerative diseases, research by our team and others has put inflammation in a different light, encouraging scientists to search for ways to harness the immune system to treat such illnesses.

Physiological inflammation is now seen as the natural product of a properly functioning immune system; this inflammation becomes chronic only if it is not regulated.

It now appears that the risk factors for Alzheimer's disease, ALS, and other chronic neurodegenerative diseases are present for much of our lives, but they are neutralized by a properly functioning immune system. They remain dormant and undetected as long as the immune system can contain them. When the immune system fails, or when immune cells are prevented from finding their way into the brain or spinal cord, these diseases take advantage of our reduced defenses.

Another undesired potential outcome of reduced immune defenses is the emergence of neurodevelopmental diseases. The brain goes through several critical points in its development, from the embryonic stage to late adolescence and early adulthood. Throughout this period the brain undergoes extensive changes, including the building of new networks between neurons, the remodeling of old ones, and the pruning of some.

This process is crucial for acquiring a wide range of sensory and cognitive functions, and requires a functioning immune system that supports these developmental changes and keeps the brain healthy during this pivotal period. It is thus likely that when the immune system fails to do so, pathologies emerge— including autism, schizophrenia, Tourette's syndrome, attention deficit disorders, and obsessive-compulsive disorders.[33]

We now know that patients with various neurodevelopmental disorders suffer from congenital immune dysfunction whose symptoms emerge at different stages after birth. In autism, for example, the unregulated immune response has been linked to disturbances in mood and sleep. Increased production of antibodies and inflammatory compounds has been suggested as a possible cause of Tourette's syndrome and obsessive-compulsive disorders, where such immune components disrupt neuronal circuits.[34]

In Rett syndrome, an autism spectrum disorder that develops early after birth, immune cells in the brain are impaired in their ability to clear broken cells and cell debris, causing neuronal dysfunction. A study conducted by Jonathan Kipnis, an alumnus of my laboratory, who is now a professor of neuroscience at the University of Virginia, showed that this miserable, fatal genetic disease was significantly arrested by the replacement of defective brain immune cells with functioning blood macrophages.[35]

Our own studies indicate that in congenital schizophrenia, the symptoms of which are manifested around puberty, patients are born with a dysfunction of the immune system, which leads to a reduced expression of molecules needed by the maturing brain. Accordingly, the molecules that normally increase around puberty are absent in the brains of mice that develop schizophrenia-like disease, leading to the development of various behavioral abnormalities.[36]

Immune system dysfunction during critical periods in brain development could interfere with the normal development of the maturing brain, eventually leading to the onset of the various neurodevelopmental symptoms. In addition, immune dysfunction may impair the body's ability to eradicate or prevent infections, which could potentially damage neural tissues during this crucial window of brain development. Overcoming immune dysfunctions may be a way to alleviate in the most comprehensive way some of the symptoms of these developmental diseases.

Our research points toward new directions for improving and lengthening the lives of patients with Alzheimer's disease and ALS by boosting the immune system in a controlled process, tailored to the disease. Such treatments are currently under development and will provide new ways to manage neurodegenerative diseases and transform them from terminal illnesses into chronic ones.

7

Males' and Females'
Different Immune Systems

M en and women are physically different. This is where the consensus ends and debate arises as to whether and to what extent these differences affect our behavior and cognitive functions. Now that we have learned that the immune system is affecting almost every aspect of our brain function, the inevitable question is whether the sex-related differences in the immune system influence the way men and women think and reason or cope with central nervous system damage.

———〰〰〰〰〰———

In a 2005 conference on women and minorities in science, Lawrence H. Summers, then president of Harvard University, ignited a firestorm when he suggested that women's disproportionately low representation in science and engineering careers could be explained by innate differences between men and women. In his controversial talk, which he later said was meant to be provocative, Summers suggested that more women than men are reluctant to put in the hours required by high-powered, intense career. In fields like engineering and science, he implied, intrinsic aptitude and genetics might better explain the underrepresentation of women than might gender-based discrimination or socialization. He gave the example of his own two daughters, who were given toy trucks at a young age and treated them like dolls, calling them "daddy" truck and "baby" truck. Summers's attempt at provocation backfired and ultimately was a factor in his resignation as the president of Harvard. His remarks nevertheless stirred an enormous amount of interest and fueled an ongoing cultural conversation about the science of gender differences, and the complexity of attributing the behavior of men and women to either innate or acquired traits.[1]

This is a fraught issue and we by no means wish to tip the scale toward either genetics or socialization. Rather, we would like to suggest that there is an additional factor involved: the gendered immune system. Men and women differ in the way their immune systems function. This may explain in part some differences in behavior, the different ways men and women cope with stressful events, and the incidence and severity of central nervous system disorders as the sexes experience them.

Even when we acknowledge that gender isn't black or white, but a spectrum of many colors, men and women do display different cognitive performance and emotional response. These differences are evident in the ways men and women communicate and use language, the ways each gender expresses fear and anxiety, and their different tendencies to develop such neurodegenerative diseases as Parkinson's, Alzheimer's, ALS, and Huntington's.[2]

Neurological disorders such as dyslexia, stuttering, attention deficit hyperactivity disorder (ADHD), and early-onset schizophrenia are more frequently diagnosed in boys than in girls. Then again, girls are more prone to develop major depression, anxiety, and panic disorders. Some neurodevelopmental disorders prevail in girls and some in boys. Why should this be?[3]

Several of these differences have been attributed to the disparities in brain development between the genders. Such disparities appear to start in the uterus. In one study researchers found that female fetuses had thicker brain tissue than male fetuses in the area that connects the two sides of the brain. This area may remain stronger in adult females, allowing activation across both hemispheres of the brain, which might explain the superiority of women in the use of language and fine motor skills. In addition, a recent study revealed that rearrangement and optimization of brain circuitry occur in girls earlier than in boys, which may explain why females outperform males in certain cognitive and emotional areas and mature faster (that is, reach certain cognitive and emotional milestones on a faster schedule) during childhood and adolescence. On the other hand, boys outperform girls on spatial tasks. Several studies have tried to explain these differences by showing that boys use more parts of their brain for spatial tasks.[4]

Is this an inborn trait determined by our sex chromosomes? Some studies have shown that testosterone affects our ability to rotate three-dimensional objects in mental space. Men appear to perform better in this area, as well as women who were exposed to excess testosterone in the uterus. In fact, a recent study found that if you are a female with a male twin (and therefore are more likely to have been exposed to a high level of testosterone in the womb), you will probably score higher on the mental rotation task than other females.[5]

Is this gap an inherited genetic feature we received from our prehistoric ancestors? Some paleoanthropologists theorize that men were hunters who traveled away from home base and had to rely on the location of the sun and their own navigation skills to find their way back. Women stayed closer to the home base, foraging, fishing, and trapping in the vicinity of the cave and therefore relying on familiar landmarks to navigate their route back. Alternatively, it may be related to the culture we grew up in. Could this be the origin of the gap between men and women? Is it nature or nurture?[6]

This question returned to the spotlight after Summers's provocation, and it occupies neuroscientists and psychologists to this day. At first, it seemed that advances in research techniques could solve the mystery. Numerous fMRI studies compared the structure and activation of male and female brains in the hope of finding a biological explanation to our gendered behavior. However, given what we know today about brain plasticity, even such biological differences may well be a result of the way our brain was shaped by society. In simple words: Your brain is what you make out of it—experience can change your brain.

There is, however, an additional factor that is evidently different between men and women and was largely ignored in this debate. Men and women are innately different in their immune system. Immune responses are sexually dimorphic—different—both in type and magnitude. In general, women have a more intense immune response than men. This might explain why the incidence and severity of human diseases varies between the sexes. While males are more prone to infections, females tend to develop autoimmune diseases in greater numbers. Does this imply that men and women are differently protected against neurodegenerative conditions? The answer appears to be yes, certainly when the onset occurs before menopause. Why should this be so?[7]

So far we have learned that the immune system can affect our ability to think and reason (Chapter 2). It dictates our mental function (Chapter 3) and the way we cope with central nervous system damage (Chapters 4 and 5). We have seen that the fitness of our immune system controls the way we age as well as the way we handle stress and trauma. Is it possible that the different immune responses men and women have are responsible for differences in behavior we observe between the two genders?

LOSING COGNITION TO PROTECT THE FETUS

We were tossing around this question while one of the students in the laboratory was pregnant and complaining about the cognitive deterioration she was experiencing. Every woman who has been pregnant can relate to this feeling. You suddenly forget where you put your keys, have trouble concentrating and remembering things. You might be suffering from "pregnesia" or "baby brain." The scientific community has tried to validate this phenomenon. Several studies found evidence of memory impairments in pregnant women compared with women who weren't pregnant. Other studies defined the cognitive changes women experience not as a negative deterioration but rather as a positive bias of the brain, focusing on cognitive skills that will boost the woman's capability to care for her vulnerable offspring. In any case, the inevitable question was what drives these cognitive changes?[8]

"Oh sure, it's your hormones." That is the most common response a pregnant woman gets when sharing her experience with people around her. But as scientists dealing with the close relationship between the brain and the immune system, my group couldn't help but wonder whether the cognitive changes my student was experiencing were related to immune system changes during pregnancy.

As the fetus develops from the fertilized egg, it contains biological elements that derive from the father and are thus foreign to the mother's immune system. You can think of the fetus as a kind of graft growing inside the woman's body. We know that when we transplant an organ, we need to make sure that the donor is well matched to the recipient. Moreover, recipients usually get powerful immune-suppressing drugs to prevent rejection of the donor organ. How is it possible, then, that a foreign tissue (the fetus) can grow undisturbed in the womb without the mother's immune system rejecting it?

In 1953 Peter Medawar, a British immunologist, asked this question in a groundbreaking lecture at a meeting of the Society for Experimental Biology. Medawar, who was awarded the Nobel Prize in 1960 for his work on the immunology of organ transplants, suggested that to tolerate the fetus for forty weeks, the mother's immune system has to be suppressed. Pregnancy induces changes in the number and function of maternal circulating immune cells. Generally, adaptive immunity (the function of cells with immunological memory) is weakened to avoid fetus rejection, while the nonspecific immune cells that provide the first line of defense, making up the innate immune system of the pregnant woman, are switched on to protect her from bacteria and viruses.[9]

When we discovered in 2006 that the immune system supports creation of new neurons in the brain, one of my graduate students told me a story about her friend who was pregnant. During the pregnancy she was diagnosed with breast cancer.

My graduate student asked me whether the emergence of her friend's cancer was related to her pregnancy. Is it possible that her pregnancy-related immune changes predisposed her body to the growth of the tumor? We discussed these immune changes and the common phenomenon of reduced memory skills during pregnancy, and all of a sudden we realized that it might all be connected to our findings! Could it be that the reduction of adaptive immunity during pregnancy as a way of avoiding fetus rejection is the cause of cognitive loss? Within five minutes, we were already designing the experiment to challenge the notion.

Our cognitive ability depends in large part on neurogenesis —creation of new brain cells, which continues throughout our life. These new cells enable us to think better, cope with stress, and store memories. When we compared pregnant mice with nonpregnant ones, we found reduced neurogenesis in the pregnant group.

As we saw in Chapter 2, we have proven with animal studies that immune cells, specifically those that are part of the adaptive immune system, the "immune cells of wisdom," are crucial to the regulation of formation of new neurons in the adult brain and of normal cognitive function. We therefore suspected that the pregnancy-induced decline in adaptive immunity might explain the prosperity of the tumor, on one hand, and the reduced formation of new neurons and the impaired cognition women experience during the nine months they are carrying a fetus, on the other hand.

If immune changes are behind this "brain fog," we should find no pregnancy-related reduction in formation of new neurons during pregnancy in the brain of immune-deficient mice. Indeed, pregnant immune-deficient mice did not show any reduction in formation of new neurons relative to virgin immune-deficient mice. Thus both groups presented low levels of new neuron formation, as both the virgin and the pregnant mice lacked immune cells that could promote generation of new neurons. However, when we placed the immune system back in the equation, by injecting the immune deficient mice with immune cells, practically restoring a normally functioning immune system, the pregnancy-induced reduction in neurogenesis became evident. Immune-deficient virgin mice, whose immune system was restored, showed increased formation of new neurons compared with their pregnant counterparts. While the virgin mice could profit from the restoration of immune cells, pregnant mice couldn't because of the immune suppression in pregnancy. We concluded that the reduced formation of new neurons and possibly the cognitive decline associated with pregnancy are related to the "price" of avoiding rejection of the fetus by the immune system.

In an attempt to find ways to temper the pregnancy-induced reduction in formation of new neurons, we placed pregnant mice in cages that allowed them to run on wheels if they wished. These pregnant mice had elevated levels of new neurons comparable with virgin mice, indicating that moderate physical activity during pregnancy might just save mothers-to-be from "pregnesia." It seems that the reduction in brain plasticity during pregnancy can be overcome by physical activity, which boosts the immune system.

Pregnant women who for whatever reason can't exercise must be patient. Our studies revealed that shortly after our pregnant mice gave birth, the immune system balance was restored and mice regained normal cognitive function (figure 17).[10]

THE GENDER OF THE INJURED CORD

What about spinal cord damage? Do men and women cope differently with assaults to the spinal cord? One of my graduate students at that time, Ehud Hauben, found that young female rats and mice recover from spinal cord injury significantly better than their male littermates. While the female rodents were able to extend their hind limbs and use them to move forward, male rodents dragged their hind limbs and remained largely paralyzed. We found that this female advantage disappeared when the spinal cords of immune-deficient male and female rats were injured, highlighting that the gender-determining effect on recovery is mediated through the immune system. In line with this understanding, when females were treated with male steroids, their advantage in postinjury recovery vanished. After treatment with steroids, female rodents performed worse on a motor task, similar to the performance we observed in the injured male rodents.[11]

Studies both in animal models and humans have shown that the male hormone testosterone suppresses the immune

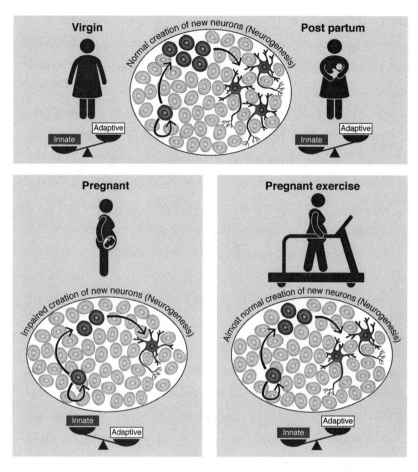

Figure 17. Immune changes during pregnancy affect the creation of new neurons (neurogenesis). The woman's immune system, which is normally biased toward a strong adaptive immunity (virgin; left upper panel), is suppressed during pregnancy to protect the fetus (pregnant; left lower panel). Such reduction in adaptive immune cells, which are crucial for brain neurogenesis and cognitive function, results in impaired creation of new nerve cells and cognitive decline during pregnancy (pregnant; left lower panel). Moderate physical activity during pregnancy boosts the adaptive immune response and results in almost normal levels of neurogenesis, which rescues the pregnant woman from "baby brain" (exercise while pregnant; right lower panel). Shortly after labor immune balance is regained, and neurogenesis levels and cognitive functions rebound (postpartum; right upper panel).

system. Castrated male rats, from which testosterone is eliminated, showed improved motor function following spinal cord injury compared with male rats that were not castrated. We also found that a drug that blocks testosterone function improved spinal cord recovery in male rats. Thus testosterone-induced immune suppression may make men more vulnerable to and less capable of coping with spinal cord injury. These results might partly explain the worse outcomes of spinal cord injury in young males compared with young females.[12]

Female advantage is also evident in other acute central nervous system injuries, as female hormones, which modulate the immune response, afford protection and support repair in brain trauma and stroke.[13]

Thus it seems that by being able to mount a strong immune response, essential for brain maintenance and repair, women can better cope with central nervous system injuries. However, this benefit comes at a price. When such a response spins out of control, women pay a penalty: a higher incidence of autoimmune disease. In fact, the prevalence of autoimmune disease in young females is higher than in young males. Since this response is dependent on hormone levels, the prevalence of most autoimmune diseases decreases with age. By the same token, women lose their gendered immunological advantage with age. This is probably due to the reduction in estrogen, a hormone that stimulates the immune system. As women enter the sixth decade of life, they see their risk of autoimmune diseases go down at the same time they lose their advantage in coping with neurodegenerative conditions. The reduction in estrogen may also partly explain the high prevalence of Alzheimer's disease that usually occurs after age sixty-five among women.[14]

Gender differences exist in the prevalence of other chronic neurodegenerative diseases as well. For example, men are more susceptible to amyotrophic lateral sclerosis (ALS) than women, who also get the disease later in life compared with men. Our own work and studies by others have shown that ALS involves an impaired immune system. Compared with men, women's immune response is generally more active and intense, especially in their childbearing years. They might have an advantage in coping with ALS risk factors. Additionally, as ALS predominates in young men, whose testosterone levels are usually high, it is tempting to speculate that these high testosterone levels, which weaken the immune response, might contribute to the development or progression of ALS. In other words, overproduction of testosterone, which suppresses the immune system, may leave young men at risk for ALS.[15]

It is clear that sex-related as well as age-related differences in the immune system powerfully affect many aspects of both normal and pathological brain functioning. This understanding highlights the importance of research into such differences. Scientists should incorporate sex and age as variables in their experimental design. Immune-related differences between the sexes also raise the value of personalized medicine that customizes health care, providing medical practice tailored to the individual patient.[16]

No matter how you put it, sex matters. Males and females are different in many aspects. We now know that at least some of these differences can be attributed to the diverse ways the immune systems of men and women function.

Epilogue

In this book, I have shared with you the research findings of my group demonstrating the broad influence of the immune system on our brain and our mind. The immune system serves as a natural antidepressant. It has the potential to keep our brains young, to reverse neurodegenerative diseases, and to cure spinal cord injuries. We have shown that the experienced immune cells help shape our ability to utilize our mental and cognitive potential. These immune cells, the "immune cells of wisdom," control the way we think and reason, the way we learn and form memories, and the way our brain copes with the constantly changing environment. These are autoimmune cells that recognize components of our brain, cells that were considered our enemies back in the 1950s and are now appreciated as key players in brain maintenance and repair.

Hippocrates' fundamental Law of Cure—"Give me fever and I can cure every disease"—suggests that fever is a reflection of nature's efforts to remove external intruders from one's system; such intruders include "morbid matter," virus, poison, or microorganisms dangerous to health and life. (Blessed efforts can sometimes act as a double-edged sword, as a fever that is too high and lasts too long can turn into a disease by itself.) In the same spirit, I view autoimmunity as the body's physiological mechanism to fight against internal intruders, physiological compounds that change their structure (misfolded proteins), composition, or levels. As with fever, an autoimmune response that is too strong for too long can turn into a disease by itself.

Equipped with this knowledge and understanding, you are now probably asking: What's in it for the patients? What's in it for me and my loved ones? Could this knowledge help to improve my memory? Could it help my mother-in-law think clearly as she struggles with Alzheimer's disease? Could it assist my wife, who recently became a mother, muddle through her anxieties and changing moods? Could it help my son, who lately returned from a long military service, regain his normal life? Could it make the girl next door walk again after being involved in a car accident that left her paralyzed from the neck down? Where do all these discoveries lead us? Will they change the way brain/mind diseases are treated ten years from now? We believe that they can in fact have all these benefits. The inevitable question is what should be done further to what has been accomplished so far to make this change a reality.

—◇◇◇◇◇◇◇◇◇—

First, we need to conduct more basic research. Many research questions remain to be explored. The leap of scientific imagination that sent me on this journey was followed by almost two decades of extensive research by generations of outstanding students and colleagues, without whom we would not be able to reach the destination. The process reminds me of a quotation from Einstein: "Imagination is more important than knowledge." I do believe that the imagination is key for such a journey, but only through ongoing state-of-the-art scientific research can we get the knowledge that can assemble the pieces of the puzzle so that we might further improve our understanding of the brain and find new ways to treat brain pathologies.

As you can appreciate from reading this book, challenging the conventional dogma wasn't easy. As often happens in science, thinking outside the box can earn one many opponents. In the early stages, because of the rampant skepticism, I asked each new student who joined my team to repeat our "historical" experiment as part of their basic training; the joy of reproducibility was my strongest scientific satisfaction at that time. Now that many such experiments have been repeated and validated by several other laboratories and my theory is increasingly accepted by the scientific community, I can honestly say that the struggle was worthwhile. A scientist who has confidence in his or her vision—and patience—has the keys to success. Notably, my scientific journey could not have been accomplished without resilience to opponents (skepticism, competitors, "the establishment"), hard work, outstanding students and colleagues who believed in the vision, and supportive governmental and nonprofit agencies.

What needs to happen to make that vision a reality? To incorporate the scientific knowledge in real practice we need to establish a new discipline in medicine, immunoneurology —a discipline that will allow us to bring what we know about the way the immune system influences the brain and mind to the daily practice and standard health care provided to patients. Medical schools should update their curricula. Immunoneurology should become a specialty, with its own medical board and residencies.

Just imagine a waiting room of a bustling immunoneurologist's practice. Over there, seated in a wheelchair, is a young woman with spinal cord injury. Beside her is a young man newly diagnosed with ALS, and next to him a middle-aged woman with early-onset Alzheimer's disease. Across the room there is an elderly man with a seeing-eye dog, hoping to have his vision restored. There is a police officer with posttraumatic stress and a new mother whose postpartum depression isn't resolving on its own. All have come to consult with the immunoneurologist and plan their treatment, harnessing their own immune systems to fight their neurological ailments.

To help immunoneurologists do their job, we should develop cutting-edge diagnostic tools based on the interaction of the brain and the immune system. For example, we could look for the optimal blood test that determines our immunological age, and the levels of our experienced "immune cells of wisdom." This test could be given routinely at annual physicals. Results of this test could give health care providers data about our immunological fitness and our vulnerability to brain disease, and it could help predict the way our body and brain could cope with such ailments. One such example is the Neuro-Quest kit, designed for measuring immunological components for diagnosis of early onset neurodegenerative diseases, currently under development based on our intensive research.

In addition to diagnostics, our studies could promote the design of personalized treatment options; we could create a personal bank of experienced T cells, comprising immune cells of wisdom that will be expanded and ready to use when needed.

The biggest challenge, of course, is bringing these potential therapies into the clinic and making them a reality for human patients. Elated as we were by our findings in the laboratory, we are well aware that many promising potential therapies do not make it to the clinic. This is due in part to enormous costs of translation that are beyond the capacity of academia, and in part to the difficulties of clinical translation from animal studies to human patients. At the translation stage, all of us in academia are dependent on drug companies or biotechnology start-up companies.

Pharmaceutical companies are businesses, and they are understandably dedicated to making a profit for their investors and shareholders. Unfortunately, this isn't always the best way to leverage promising animal studies into therapies that can transform patients' lives. While drug companies have the resources to move ideas forward into the clinic, some of them license out their intellectual property, but eventually shut the drug development process off, due to financial conflict of interest. In such a case, much money is spent on legal issues rather than on advancing the drug, a situation that keeps many promising therapies in the laboratory and out of the clinic. This occurred in my case with some of the innovative therapies that could have been tested by now for treatment of Alzheimer's disease and glaucoma. Our results in animal models have shown us that our approach could reverse these diseases. If the dispute with the drug company can be resolved, the way will be paved to test these drugs as a potential therapy for some of the most devastating diseases of the modern world.

So what is being done now?

Taking the scientific journey a step farther, we envisioned several potential immune-based therapies. One is a therapeutic vaccination for Alzheimer's disease, ALS, glaucoma, and depression. Another is a drug that reduces the activity of suppressor cells, which restrict the protective immune response. A third is manipulating the immune system at the brain's border as an alternative therapeutic strategy for restoring brain function in aging and in Alzheimer's disease. Another possible treatment might restore the pool of "immune cells of wisdom" to prevent the cognitive decline that occurs in aging.

Our therapeutic strategy in the most advanced stages of development is the transplantation of immune cells (macrophages), a cell-based therapy for spinal cord injury patients. Phase I clinical trials of this experimental immune cell–based therapy were conducted successfully from 2000 to 2003 in Israel and Belgium. Of the eight patients participating, three recovered clinically significant neurological motor and sensory function. A biotech company, ImmunoBrain Therapies, is currently developing a new and improved protocol for phase II clinical trials.

The challenge has been to find the optimal way to translate our know-how from mice to humans, to choose the human cells that will be the closest counterpart of the mouse cells, or to choose a suitable therapeutic vaccine for humans that will be the equivalent of the one shown to be effective in the mouse.

By providing safe and effective immune-based therapies that will help the brain to cure itself, we hope that our advanced understanding will one day fight off chronic neurodegenerative diseases (ALS, Parkinson's disease, Alzheimer's disease, glaucoma, and others), reverse mental disorders (depression and posttraumatic stress disorder), help one to maintain a sharp mind throughout the entire lifespan, and halt aging in the brain. It is my hope that we won't have to wait a generation to begin to see the promise of immunoneurology reflected in patient lives.

A NEUROIMMUNOLOGY PRIMER

THE CENTRAL NERVOUS SYSTEM

The central nervous system (CNS) is the major system of your body that determines who you are and how you cope with the environment around you, gathering up sensations and organizing them into your waking, conscious experience of the world.

The Neuron

The basic working unit of your central nervous system is the neuron, a nerve cell designed to react to stimuli and deliver information in the form of an electrochemical signal to other

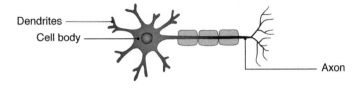

Figure 18. Classical central nervous system neuron.

cells in the body (figure 18). The human brain includes between 100 billion and 200 billion neurons interconnected in a complex network among themselves and with other cell types.

The neuron consists of a cell body, containing genetic information and responsible for the cell's metabolism; dendrites, which extend from the cell body like antennae, collecting information and transmitting it to the cell body; and the axon, a nerve fiber that acts as an electrical cable, conducting information to other target cells (neurons, muscles, or glands), telling them what to do or make.

Every function of your brain depends on such electrical transmissions. In fact, when you are awake your brain can generate just enough energy to power a small light bulb, around twenty watts.

Just as electrical wires in your home are covered in thermoplastic insulation, your axons too are covered in protective material. Named myelin, this insulation helps messages travel along the axon more quickly, and makes the delivery of information more efficient.[1]

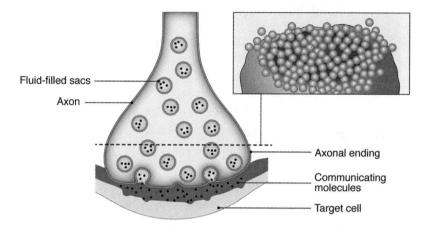

Figure 19. Transport of neurotransmitters (brain-derived communicating molecules) from the end of the axon to the target cell. Box to the upper right shows a nerve broken open to reveal the vesicles (fluid-filled small sacs) containing neurotransmitters.

Neurotransmitters

Once the electrical impulse arrives at the target cell, it is delivered chemically, by the release of certain molecules from small sacs (vesicles) at the end of the axon. These molecules are called neurotransmitters (figure 19). They are used for communication within the brain, and between the brain and the rest of the body. They control our moods, weight, sleep, and energy, and many other aspects of our daily experience.

One key neurotransmitter is serotonin, also known as one of the chemicals of "happiness" that our body produces, which helps people to cope with short- and long-term stress. This is the "don't panic" neurotransmitter, controlling our emotions, calming anxiety, and relieving depression. Serotonin regulates your appetite and body temperature and is responsible for your good night's sleep. It even controls a mother's milk supply while she is breast-feeding. Several drugs used to treat depression, including fluoxetine (Prozac) and sertraline (Zoloft), increase the availability of this neurotransmitter and hence provide an antidepressive effect.

Another important neurotransmitter is dopamine, a molecule that affects our reward and pleasure systems, motivating us to perform certain acts of behavior. Drugs such as cocaine, amphetamines, and alcohol augment the dopamine effect. They increase the levels of dopamine in your brain, which over time desensitizes your neurons so they do not "feel" that they are getting enough dopamine anymore. Your brain associates the drug with the neurochemical reward, and the urge to take more of the drug increases. Dopamine also regulates our movements, and a deficiency in this neurotransmitter is considered the key cause of Parkinson's disease, often diagnosed with the onset of a tremor or problems with movement. Erratic dopamine activity is also evident in patients with schizophrenia and in people with attention deficit/hyperactivity disorder (ADHD).

Another neurotransmitter, key to all the brain's activity, is glutamate. Glutamate is involved in most aspects of normal brain function, including cognition, memory, and learning. Yet glutamate has to be present in the right concentrations in the right places at the right time. Either too little or too much glutamate can be harmful. Excessive levels can damage the brain and lead to a range of neurodegenerative diseases. Many attempts have been made to develop drugs that balance glutamate levels as potential treatments for these diseases.[2]

The Brain

Our large brain sets us apart from all other species. This three-pound spongy mass controls and coordinates every aspect of your life: every breath, every beat of your heart, the way you learn and store memories, every emotion and feeling, every sex drive, to the way you move, work, and play.

The brain is made up of many parts, all wired together to form a highly complex, multitasking organ. It is this complexity that makes this organ unique. It is more than a collection of axons. There is no man-made machine equivalent to the human brain. This is why computers are far from being able to replace the human brain and mind. Some futurists believe they never will, and so do I.[3]

The Spinal Cord

The brain extends from the brain stem through the vertebral column to form the spinal cord, a column of nerves that connect your brain with the rest of your body. The spinal cord is encased in the bony vertebrae that make up the spine, or backbone. This segmentally arranged structure runs from the base of your skull to the pelvis. In addition to protecting your spinal nerves, it supports your body's weight, and allows you to bend and flex your torso and maintain an upright posture. In men, the spinal cord is about eighteen inches long, in women, about seventeen inches. This is about the same as the distance from your hip to your knee. The cord is as thick as your thumb throughout most of its length.

Each segment of the spine corresponds to spinal nerves that control a certain body region. When it exits the vertebra, the spinal nerve splits so that some branches turn to the front part of the spine, toward your chest, while others turn to your back. The nerve branches that extend to the front of the body carry information from the brain to muscles, glands, and internal organs, while the ones that enter at the back of the spine relay sensory information from the body regarding its position in space, as well as temperature, pain, pressure, and touch. All this information is routed to the appropriate regions in the brain. This might explain why the traditional practice of massage provides relief; the massage therapist is applying pressure where the sensory pathways are concentrated.

Following injury to the spinal cord, the brain loses its ability to communicate with parts of the body controlled by spinal nerves located below the level of the injury. The injured person may eventually lose function of these body parts. Usually the closer to the head the spinal cord injury is, the greater the area of the body that may be affected, sometimes resulting in complete paralysis.[4]

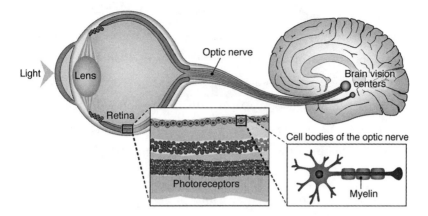

Figure 20. The retina as an integral part of the central nervous system. At the back of the eyeball lies the neural retina, which consists of cells that collect the light (photoreceptors) and cell bodies of the optic nerve that deliver the information to the brain (retinal ganglion cells) (middle box). The retinal nerve cell that makes up the optic nerve has the classical structure of a central nervous system neuron (right box). The axons of these cells are collected in a bundle that forms the optic nerve, which extends to brain vision centers, which interpret what we see. Figure is modified from London, Benhar, and Schwartz, "The Retina as a Window to the Brain," used with permission of *Nat Rev Neurol.*

The Neural Retina

An additional part of the central nervous system is the neural retina, located in the back of the globe of the eye (figure 20). The process of vision begins when light enters the eye through the cornea, a transparent membrane that acts as a window and covers the front part of the eye. The light then passes through the lens, which adjusts the focus, finally projecting a clear image onto the retina.

The retina consists of several layers. The outer layer contains photoreceptors, nerve cells that absorb light and convert it into an electrical signal, which is transmitted to the innermost layer of neurons, called retinal ganglion cells, the cell bodies of the optic nerve. The axons of these retinal ganglion cells are gathered in a bundle, forming the optic nerve that extends to the visual centers in the brain, where the visual information is processed and interpreted. These optic axons are covered by myelin and are in fact ordinary central nervous system neurons.

The front of the eyeball contains a cavity filled with a clear fluid called aqueous humor that nourishes the lens and cornea. Defective drainage of this fluid leads to increased pressure inside the eye, damaging the nerve cells in the retina, and initiating a self-perpetuating process (a "domino effect") of cell death associated with an eye disorder we know as glaucoma.[5]

The Blood-Brain Barrier

The central nervous system must be protected from physical shocks and pathogens and must have an adequate supply of nutrients to ensure its sophisticated function as the body's main "control" room, orchestrating the function of all other organs and systems in our body, and facilitating the appropriate responses to a constantly changing environment. Moreover, since the communication among neurons involves delicate ionic circuits, the brain needs to be protected from ions leaking from the circulating blood. Although central nervous system tissues require a constant blood supply, there is no direct contact between the neuronal tissue and the blood vessels. Instead, central nervous system tissues are separated from the blood circulation by a unique structure covering all the blood vessels that feed the neuronal tissue, called the blood-brain barrier. In this interface between blood vessels and central nervous system, the cells that form the blood vessels (endothelial cells) are tightly packed, preventing ions, large molecules, most proteins, and cells from passing into the central nervous system.

Although the brain makes up only 2 percent of the weight of the average adult, it is one of the biggest consumers of oxygen in the body, consuming 20 percent of the oxygen in every breath you take. Luckily, oxygen can easily pass through your blood-brain barrier. Most of that oxygen is used by the nerve cells in the central nervous system to break down sugars, mostly glucose, in a process that releases energy. Glucose enters the central nervous system through a special carrier embedded within the blood-brain barrier. The brain's constant demand for oxygen is the reason why having a stroke is so dangerous. The brain cannot store oxygen and relies on blood vessels to continuously supply blood rich in oxygen. When one of these vessels becomes blocked by a blood clot, the nerve cells in that area of the brain don't get their share of oxygen. These cells die within minutes. This can result in severe brain damage and even death.

If the central nervous system becomes impaired by disease or injury, some of the properties of the blood-brain barrier are lost, resulting in an imbalanced environment. Immune cells flood the neural tissue, worsening the disease or damage. For instance, in multiple sclerosis (MS), the cells that form the barrier undergo changes that facilitate uncontrolled entry into the central nervous system of immune cells that jeopardize nearby neurons.

While it plays an important role by excluding harmful substances from the central nervous system, the blood-brain barrier can also prevent therapeutic drugs from reaching the brain. In fact, finding ways to get drugs into the central nervous system is one of the toughest challenges facing researchers in the pharmaceutical sciences. A breakthrough in drug delivery to the central nervous system could change and even save the lives of many patients suffering from neurological defects, brain diseases, and brain cancers.

One substance that can easily pass through the blood-brain barrier (although often we wish it couldn't) is alcohol. Small amounts of alcohol first affect areas responsible for inhibition; this explains the euphoric "high" feeling and greater sociability many drinkers experience. Unfortunately, it also explains the slowed reaction time, impaired vision, vague thinking, and memory loss, as alcohol reaches neurons and delays their function. In addition, several studies have shown that heavy drinking impairs the blood-brain barrier, thereby exposing the individual to various central nervous system pathologies, such as neuroinflammation, stroke, Alzheimer's disease, and dementia.

With that in mind, you will be happy to learn that not all drugs are bad for your brain. Scientists have recently suggested that a cup of coffee a day might keep Alzheimer's away. In this study, caffeine, one of the world's most popular stimulant drugs and another substance that easily crosses the blood-brain barrier, was found to stabilize the barrier and prevent the kind of barrier dysfunction associated with brain disorders such as Alzheimer's. Although key to the stable environment of the neuronal tissue, the blood-brain barrier is not the sole interface between the brain and the circulation. There are other crossing points in which immune cells and brain tissue meet.[6]

The Meninges and Cerebrospinal Fluid

The central nervous system is ensheathed by three layers of connective tissue, collectively termed the meninges. The meninges contain the cerebrospinal fluid, which flows between their layers, and support the blood vessels that feed the brain. These structures work together to protect and nourish central nervous system tissues. Meningitis is a viral or bacterial infection of the meninges. Meningitis can lead to such complications as deafness, brain damage, and learning disabilities, and can even be life-threatening.

Covered in the meninges, the brain floats in cerebrospinal fluid (figure 21). This crystal-clear liquid acts as a shock absorber by providing a "fluid blanket" surrounding the nerve tissues. Cerebrospinal fluid also plays an important role in brain metabolism, removing waste from the neural tissue and

183

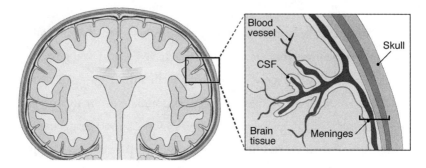

Figure 21. The cerebrospinal fluid and the meninges: The brain floats in the cerebrospinal fluid. Enlarged box: The cerebrospinal fluid (CSF) flows between two of the layers of the meninges (layers of connective tissues that shield the brain) and covers brain tissue. Figure is modified from Shechter, London, and Schwartz, "Orchestrated Leukocyte Recruitment to Immune-Privileged Sites," used with permission of *Nat Rev Immunol.*

supplying oxygen and nutrients from the blood. Your bloodstream delivers oxygen and carries away carbon dioxide, enriching the cerebrospinal fluid with water and such substances as sugar, several kinds of white blood cells, and different ions.

In a healthy person, the cerebrospinal fluid lacks any blood components that could interfere with the delicate electrical activity of the brain. This allows the central nervous system to benefit from certain blood-derived supplements while maintaining the chemical balance it needs to function. Doctors long believed that this "clear" fluid does not contain any immune cells. Any that were present were considered the first sign of autoimmune disease. As we make clear in the book, as our understanding of the brain evolved, we came to see these immune cells in a different way. We now view some of them as essential immune cells located within the territory of the brain but outside the brain neuronal tissue itself. From the cerebrospinal fluid, these immune cells perform special surveillance of the brain, a form of immune "remote control."

When the central nervous system is damaged, the composition of the cerebrospinal fluid dramatically changes, making it useful for diagnosing certain diseases. Doctors can sample the cerebrospinal fluid via a lumbar puncture, commonly known as a spinal tap. In this procedure, a doctor inserts a needle into the space between two vertebrae in the patient's lower back. In the past, "cloudy" fluid indicated a diseased central nervous system. With our new understanding of the relationships between the brain and the immune system, we need to refine our idea of what it means to find immune cells in the cerebrospinal fluid. This may dramatically change the way clinicians interpret the analysis of cerebrospinal fluid samples from lumbar punctures. The new insights addressed in this book will thus influence not only the ability to maintain our brain in good condition but also future diagnosis and therapy for fatal diseases that are as yet incurable.[7]

Choroid Plexus

The cerebrospinal fluid is produced by the choroid plexus, a continuous bushlike structure of capillaries and specialized cells that floats in four hollow spaces in your brain, called ventricles (figure 22). Weighing only a fraction of an ounce, this tiny, highly tangled structure secretes just over a pint of cer-

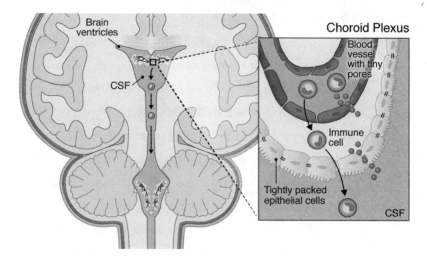

Figure 22. Brain-immune communication at the brain's border, the choroid plexus: The choroid plexus is located at the walls of the brain ventricles (cavities that exist in the brain). Enlarged box to the right shows a blood vessel, containing tiny pores, and tightly packed epithelial cells at the choroid plexus. Under normal conditions, immune cells control daily brain functions by remote control, secreting molecules to the cerebrospinal fluid, which delivers them to the brain tissue. When required, this border becomes an active gate that selects and shapes the immune cells that enter the brain to heal and repair damage. Figure is modified from Shechter, Miller, Yovel, et al., "Recruitment of Beneficial M2 Macrophages to Injured Spinal Cord," used with permission of *Immunity*.

ebrospinal fluid each day. As opposed to the rest of the brain, the cells composing the capillaries at the choroid plexus (endothelial cells) are not tightly packed. These capillaries are fenestrated, meaning they contain tiny pores that allow the passage of small molecules and proteins. Instead, a distinct barrier is formed at the choroid plexus by tightly packed epithelial cells, cells which cover surfaces of internal tissues such as the lung and intestine (figure 23).

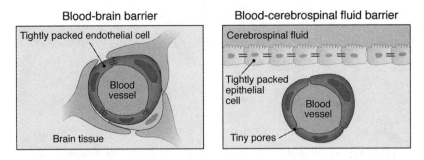

Figure 23. Blood vessels at the blood-brain barrier and the blood-cerebrospinal fluid barrier: Most of the blood vessels in the brain are composed of *tightly packed endothelial* cells (which form the blood vessels), enwrapped by supportive cells and structures that together form the highly restricted blood-brain barrier, left. Blood vessels at the choroid plexus contain tiny pores, and the barrier is based on the *tightly packed epithelial* cells (specialized cells lining internal tissues and cavities of the body).

One of my team's key discoveries has revealed this interface in a new light. We showed that this unique barrier between the brain's protective liquid, the cerebrospinal fluid, and the blood vessels functions as the site of a lifelong dialogue between the brain and the immune system. It serves as a remote platform from which immune cells can communicate with the healthy brain, assisting in its routine maintenance without actually entering the tissues. Under certain conditions, this site acts as an active gate that meticulously selects and shapes the immune cells that enter the brain. As opposed to the blood-brain barrier, which should remain sealed to immune cells under all conditions, and which, once breached, may potentially result in a pathological inflammation, we found that entry of immune cells through the choroid plexus, the brain-immune communication border, is essential for healing, provided that the structure maintains its integrity.[8]

THE IMMUNE SYSTEM

The immune system is one of the most indispensable systems of our body, functioning to keep us healthy. For the greater part of the twentieth century, immunologists viewed it as a system that solely evolved to protect us against invading pathogens, such as bacteria or viruses, but over the past few decades the immune system has increasingly been recognized for its essential role in maintenance and repair. More recently, researchers have attributed the immune system a role in coping with *internal* enemies, such as toxic molecules produced by our own bodies, or coping with our own cells if they become cancerous. Our body is replete with these enemies and other molecules that switch them on. It is the role of the immune system to promptly recognize these internal enemies and neutralize them. Perhaps with this new knowledge of the immune system, we could define it as the body's system of defense against internal and external threats.

Usually, the process of immune defense is successful, beginning and ending without us even being aware of the battle taking place inside us. But sometimes the immune system fails to eliminate the threat. At other times, even though it has eliminated the threat, the immune response is not shut off promptly, and may even enter a vicious cycle that not only does not end in a cure but can lead to the body attacking itself. Then we get sick, and can even develop autoimmune disease, cancer, or chronic neurodegenerative disease.

Most of what we know about the function of the immune system comes from observing how the body defends itself against microorganisms. Scientists in recent decades have adopted these lessons to understand and thereby improve wound healing and cancer treatment. As we show, these same mechanisms can be used to fight off neurodegenerative diseases.

Bone Marrow

This is a soft tissue primarily found in the hollow interior of flat bones like your ribs, pelvis, and breastbone. This red marrow is the primary factory that produces immune cells. There, all red blood cells and most white blood cells are born and mature, ready to meet a threatening microorganism or any other pathogenic compound and defend your body. These newly minted immune cells exit the marrow through tiny holes in your bones called foramina. Then they enter the bloodstream. Now they are ready to respond to chemical messages from any parts of the body under attack. In such cases, the tissue in trouble signals to your immune system via a chemical "S.O.S." signal carried in your bloodstream. These chemical signals recruit immune cells to the blood vessels adjacent to the tissue. Then these immune responders pass through the wall of the blood vessel, entering the tissue to attack the pathogen.[9]

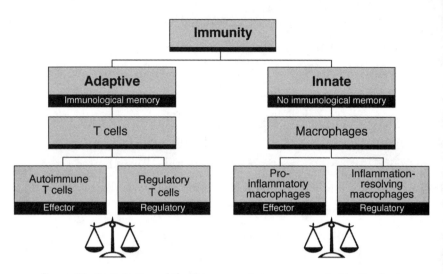

Figure 24. Organization of the immune system as represented in the book. The immune system can be largely divided into two main branches: the innate (inborn) and the adaptive (learned, acquired). The innate arm is your first line of immune defense. It is immediate and nonspecific, and does not rely on past exposure (no immunological memory). The adaptive arm is delayed, highly specific to a certain compound (antigen), and based on a successful past recognition of the antigen (immunological memory). In this book we have focused mainly on macrophages, which are part of the innate arm, and include both pro-inflammatory effector macrophages and inflammation-resolving macrophages, which regulate the immune response. On the adaptive arm we have addressed mostly T cells that largely consist of effector T cells and regulatory ones. In order to maintain a proper immune response, both innate and adaptive regulatory and effector cells must reach equilibrium.

For simplicity, we will summarize below the two main branches of the immune system, the innate (inborn) and the adaptive (learned, acquired) (figure 24).

Innate Immunity

When an invading pathogen first enters your body, it encounters several features of the human body designed to capture it. First, there is your skin, the largest organ in your body, an envelope that is your first line of defense against pathogens. Should you breathe in or swallow a bacterium, it may first encounter antimicrobial substances contained in your saliva and in the mucous membranes that line your nose and throat. If the invader makes it past these defenses, your stomach juices, which are highly acidic, having a pH between 1 and 2, await. These defenses act in the same manner regardless of the identity of the invader, or antigen.

If the threat slips past these barriers, it activates your body's innate immune cells. These are white blood cells that engulf and digest the intruder or secrete proteins that damage it. These cells act when they recognize certain patterns that are shared by many pathogens and are not specific to the intruder. Thus the innate branch of your immune system has no immunological "memory." It does not rely on past exposure to "remember" a pathogen. Should the same intruder attack again, it will be treated as if the body has just encountered it for the first time. This arm of the immune system does the job each time that it encounters an enemy regardless of its identity; the task is to remove it before it invades the tissues. It is this branch of the immune system whose role we now recognize in healing of wounds and in fighting against the body's internal enemies, dangerous compounds that are generated by our own cells and need to be removed, or those that exceed their normal levels and cause tissue damage.[10]

Adaptive Immunity

In contrast to the innate immune system, the adaptive immune branch (also called learned or acquired immunity) is highly specific. Each of its functions requires the recognition of the specific compound of a given antigen.

Cells from the innate immune branch are first to arrive on the scene. Mostly made up of macrophages and dendritic cells, they engulf and digest the threatening bacterium or virus (figure 25). Each innate immune cell retains particles of the engulfed invader on the outer surface of its cell body. These particles act as antigens, which can be recognized by the cells of the acquired immune system.

When these cells from the acquired immune system were first discovered in the late 1950s, most scientists considered them scavengers of pathogens. Now, half a century later, we know that the role of these cells goes far beyond removing intruders. These cells patrol the body, sensing any injury and facilitating repair. They are called lymphocytes.[11]

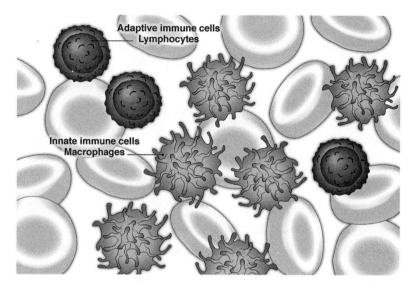

Figure 25. Innate (macrophages) and adaptive (T cells, lymphocytes) immune cells.

Lymphocytes

Lymphocytes are diverse both in the role they play in the immune system and in how they carry that role out (see figure 25). There are two major types of lymphocytes, B lymphocytes and T lymphocytes. B lymphocytes secrete antibodies, Y-shaped proteins that specifically bind to a certain antigen, eventually leading to its destruction. T lymphocytes release immune molecules called cytokines, which allow immune cells to communicate with one another. These cytokines are the language of immunity. Each has a different function; they can either take direct action toward the antigen or summon and recruit other immune cells. They can activate or suppress and regulate the overall immune response. We now know that cytokines also serve for intersystem communications; neurons can recognize some of them and respond to them. Intriguingly, it goes both ways: The immune cells can also respond to the language of the nervous system, the neurotransmitters, brain-derived communicating compounds. Thus these chemical languages serve for inter- and intrasystem communications. T lymphocytes can be either killers, directly involved in removing the threat, or helpers, assisting other immune cells in performing their roles. In this book we focus mostly on helper T cells.

The high specificity of the acquired immune response has a price—it is time consuming. It can take a few days until T cells and B cells take full action toward an antigen encountered by the body. (This is, for example, why it takes up to two weeks after getting a flu shot for you to enjoy maximum protection.) However, once this action is completed, immunological "memory" is established. Your immune system has adapted. It retains certain elements of the acquired response, such as T cells and antibodies, so that the next time you encounter the same antigen, your immune system will respond faster and stronger than it did at the first encounter.

This immunological memory serves as the basis for vaccines that prevent or treat disease. Vaccines to prevent disease contain weakened or dead pathogens or particles of pathogens. When you get a shot, these antigens are too weak to make you sick but still strong enough to drive your immune system to "remember" them and produce antibodies. In this way, when the actual pathogen enters your body, your immune system is primed to respond, efficiently eliminating the threat.[12]

In the past, vaccines were used solely to prevent illness, a way of rendering the body resilient to future pathogens, should it encounter them. We now know that vaccinations can also function as therapy. In such cases, it is not a foreign pathogen that has to be eradicated, but the enemies inside our body that have been dormant for a long time and suddenly emerge due to some loss of regulation, resulting in the appearance of a chronic disease or cancer. In such a case a vaccine can be given as therapy, helping to boost the immune system in order to arrest the disease or reverse it. Such an approach has already been adopted for the treatment of some cancers. Based on my "protective autoimmunity" theory, such vaccines hold promise in treating depression, chronic neurodegenerative diseases, and even the ravages of aging.[13]

Autoimmune Cells

But what if the antigen isn't an external invading enemy? What if the enemy is internal, made by our own body? Should our immune cells respond to such so-called self-elements? Should all self-elements escape from immune response? Alternatively, is there a purpose for immune cells that recognize self-elements?

If so, can we recruit these cells to fight against internal ene-mies without developing autoimmune diseases?

The highly specific nature of the adaptive immune sys-tem means the immune system requires an enormous range of lymphocytes, one for each of the thousands of pathogens that might enter our bodies. Scientists long wondered how this "library of lymphocytes" was created. Is it prearranged or does the specific lymphocyte form on the spot, upon demand? In 1960 the Australian virologist Frank Macfarlane Burnet was awarded the Nobel Prize in physiology and medicine for his work in the 1950s on the theory of "clonal selection."

According to Burnet's theory, we are born with the ability to create an array of lymphocytes that covers every antigen we would ever encounter. During the development of these cells, each undergoes a process of negative and positive selection in specialized organs called lymphoid organs. This process en-sures, on one hand, that the lymphocytes are capable of devel-oping into cells that will be able to recognize foreign antigens (positive selection), and on the other hand, that those lympho-cytes that might respond in an aggressive way to the body's own tissues are eliminated (negative selection). A cell that survives the selection process and eventually encounters its specific antigen is turned on and starts dividing, making more and more of the same lymphocyte to fight off the threat. That is why your immune system is stronger and more efficient the second time an antigen is encountered—it has now adapted and is equipped with lots of these specific lymphocytes, ready to act.[14]

Clonal selection theory led scientists to believe for decades that all immune cells that recognize and bind to self-components are destroyed during development, preventing the body from attacking itself. According to this view, any existing immune cells that recognize the self were considered the result of an escape from deletion, an error that occurred during the clonal selection process, and were viewed as the hallmark of autoimmune disease.

Yet during the six decades since Burnet formulated his theory of clonal selection, scientists have identified autoimmune cells in healthy individuals. A vigorous debate took place as competing theories were posited to explain why this should be so. What could these cells be doing? Have they slipped away and escaped deletion during the clonal selection process, or are they a result of a purposeful positive selection, an "intentional decision" by the immune system to have these kinds of autoimmune cells? If they do play a purposeful role, what is it?

Protective Autoimmunity Theory

My team at the Weizmann Institute in Rehovot, Israel, pioneered the theory of protective autoimmunity, suggesting that autoimmune cells are selected during the development of the immune system in a purposeful process that equips the body with a mechanism to fight off internal intruders, emerging following injury, stress, or any time the body is out of balance. Autoimmune diseases emerge not because of the presence of autoimmune cells in an individual's body but only when such cells spiral out of control once they are activated.

For instance, after an injury to the central nervous system, or onset of central nervous system disease, certain self-proteins are released into the lymphoid organs. We found autoimmune T cells that recognize these central nervous system self-proteins and prevent the chain-reaction death of neurons. This led us to revisit the perception of autoimmunity. Could these T cells, despite having a classic autoimmune profile, actually be protecting the central nervous system and helping it heal? Through a series of experiments, we showed that this was indeed the case. This protective autoimmunity goes beyond promoting healing; it provides the brain with basic mechanisms of self-maintenance.[15]

Regulatory Immune Cells

But if these autoimmune cells are allowed to circulate freely in the body, what prevents them from attacking our own tissues and inducing autoimmune diseases such as multiple sclerosis and myasthenia gravis? In addition to the classical immune cells that vigorously defend the body, collectively termed effector cells, we also have immune cells that are responsible for turning *off* the immune response. These are the regulatory immune cells. These cells exist both in the innate branch of the immune system as inflammation-resolving macrophages and in the adaptive branch as regulatory T cells. Both types of regulatory cells keep the immune response in check, intervening whenever it is too strong or fails to shut off. In a well-functioning immune system, effector and regulatory immune cells operate in equilibrium. It is when this balance is broken that diseases emerge. When there are insufficient regulatory cells, the result is chronic inflammation or autoimmune disease. This side of the scale is well accepted. But what happens if there are too many of these suppressor cells? Researchers have only recently started to consider this possibility, at least in tumors. Now it is evident that the immune system constantly eliminates aberrant cells and tumors before they initiate disease. Tumor-specific autoimmune cells are an important part of this fight against tumors. If there is an excess of suppressor cells, as is unfortunately often the case in some tumors, there will not be enough effector immune cells to eradicate the tumor. We now propose that neurodegenerative diseases, like tumors, similarly emerge when there is an excess rather than shortage of immune suppressor cells. An overrestriction of the immune system is as bad as letting it get out of control. Both are bad, but each leads to distinct diseases.[16]

Inflammation

Any immune response is accompanied by local changes in the tissue being targeted. These include swelling and redness as blood flows into the area, as well as pain and heat. These phenomena, which we all occasionally experience firsthand, are part of the process by which immune cells are recruited to help at the site of a wound or injury, and are collectively known as inflammation.

More than a hundred years ago, Russian biologist Élie Metchnikoff defined this process as a "physiological inflammation," his term for all the body's activities that strive to restore balance following any disruption. Metchnikoff went against the common wisdom at the time by emphasizing the beneficial aspects of inflammation, pointing out the key roles inflammation plays in fighting off pathogens, repairing tissue, and maintaining the body.[17]

However, over the years inflammation became almost synonymous with pathology, blamed for the major epidemics of our century, including aging and obesity, depression and cancer, and even impotence. We are often advised by the popular health media to adopt an anti-inflammatory diet, to take anti-inflammatory supplements, and even enjoy anti-inflammatory aromatherapy or massage. How did inflammation become the villain? Have we forgotten Metchnikoff's insights?

Apparently so. When most people think about inflammation, they are actually referring to pathological inflammation rather than the physiological version Metchnikoff was writing about. When the healing physiological inflammation is not resolved on time it can become pathological. Temporary inflammation develops into a chronic immune response that nurtures the development of many of our modern health epidemics. This might explain why the reputation of inflammation, a pivotal physiological response, has become entirely negative over the years, almost a synonym for a devastating pathology that the body should prevent.

Inflammation has been blamed for the major neurodegenerative disorders of the central nervous system: the disabilities associated with spinal cord injury and head trauma, depression, or any chronic neurodegenerative conditions such as Alzheimer's disease, Parkinson's disease, ALS, or age-related dementia. In this context, it may also explain why our theory, suggesting and then proving that immune cells protect the brain, was so revolutionary. Clinical trials support our view. Their results show that anti-inflammatory drugs fail to treat neurodegenerative disorders. When using a drug that indiscriminately shuts down the entire immune response, you lose the physiological inflammation essential for healing.[18]

———∿∿∿∿∿∿∿∿∿∿———

We've reviewed for you the major players in the central nervous and immune systems, focusing on those players that are relevant to the book. We hope this information can help you explore with us the new and fascinating discoveries of the brain-immune interaction, to see how our brain benefits from the immune system, and to learn about the ways we can harness the immune system to make us smarter, more youthful, and more capable of fighting brain diseases.

NOTES

1. A NEW PLAYER IN THE BODY-MIND CONNECTION

1. Clarke, *Descartes;* Glimcher, "René Descartes"; Descartes and Cottingham, *Meditations on First Philosophy;* Hatfield, "René Descartes"; Lokhorst, "Descartes and the Pineal Gland"; Schatz, "History of Neuropsychology."

2. Mangels, "History of Neuroscience"; Kandel, Schwartz, and Jessell, *Principles of Neural Science;* Schatz, "History of Neuropsychology."

3. Mangels, "History of Neuroscience"; Schatz, "History of Neuropsychology."

4. Glimcher, "René Descartes."

5. Mangels, "History of Neuroscience."

6. Northoff, "Psychoanalysis and the Brain."

7. Mangels, "History of Neuroscience"; Kandel, Schwartz, and Jessell, *Principles of Neural Science.*

8. Kandel, Schwartz, and Jessell, *Principles of Neural Science.*

9. Greenberg, "A Concise History of Immunology."

10. Shirai, "On the Transplantation of the Rat Sarcoma in Adult Heterogenous Animals"; Murphy and Sturm, "Conditions Determining the Transplantability of Tissues in the Brain"; Niederkorn, "See No Evil."

11. Shechter, London, and Schwartz, "Orchestrated Leukocyte Recruitment to Immune-Privileged Sites"; Ziv et al., "Immune Cells Contribute to the Maintenance of Neurogenesis"; Lewitus and Schwartz, "Behavioral Immunization."

2. COGNITION AND THE AGING BRAIN

1. Ramón y Cajal, *Degeneration and Regeneration of the Nervous System;* Lledo, Alonso, and Grubb, "Adult Neurogenesis and Functional Plasticity in Neuronal Circuits."

2. Altman, "Are New Neurons Formed in the Brains of Adult Mammals?"; Gould et al., "Neurogenesis in the Neocortex of Adult Primates"; Eriksson et al., "Neurogenesis in the Adult Human Hippocampus"; Kaplan, "Neurogenesis in the 3-Month-Old Rat Visual Cortex"; Gould et al., "Learning Enhances Adult Neurogenesis in the Hippocampal Formation"; Jin et al., "Increased Hippocampal Neurogenesis in Alzheimer's Disease"; Gould et al., "Proliferation of Granule Cell Precursors."

3. Kempermann, Kuhn, and Gage, "More Hippocampal Neurons in Adult Mice Living in an Enriched Environment."

4. Ziv et al., "Immune Cells Contribute to the Maintenance of Neurogenesis."

5. Ibid.; Kipnis et al., "T Cell Deficiency Leads to Cognitive Dysfunction"; Cardon et al., "Dysregulation of Kisspeptin and Neurogenesis at Adolescence."

6. Moalem et al., "Autoimmune T Cells Protect Neurons from Secondary Degeneration."

7. "Sir Frank Macfarlane Burnet"; Ziv et al., "Immune Cells Contribute to the Maintenance of Neurogenesis."

8. Voss, "The Risks of Anti-Aging Medicine"; Weintraub, *Selling the Fountain of Youth*.

9. Saltsman, "Chapter 5."

10. "Postmenopausal Hormones"; MGH Center for Women's Mental Health, "Hormone Replacement Therapy Revisited"; Beral et al., "Breast Cancer and Hormone-Replacement Therapy."

11. Fontana et al., "Long-Term Calorie Restriction"; Meyer et al., "Long-Term Caloric Restriction"; Omodei and Fontana, "Calorie Restriction and Prevention of Age-Associated Chronic Disease"; Willcox et al., "Caloric Restriction, the Traditional Okinawan Diet, and Healthy Aging"; Corliss and Lemonick, "How to Live to Be 100"; Okinawa Centenarian Study.

12. Bernardes de Jesus et al., "Telomerase Gene Therapy in Adult and Old Mice"; Toftgård, "Maintenance of Chromosomes by Telomeres and the Enzyme Telomerase."

13. Zimmer, "Gene Therapy Emerges from Disgrace."

14. Ho, Wagner, and Mahlknecht, "Stem Cells and Ageing"; "What Are Some Risks of Stem Cell Therapies?"; National Institutes of Health, "Stem Cell Basics."

15. "Henry Ford"; National Institute on Aging, "The Changing Brain in Healthy Aging"; Thies, Bleiler, and Alzheimer's Association, "2013 Alzheimer's Disease Facts and Figures"; Parkinson's Disease Foundation, "Understanding Parkinson's."

16. Lages et al., "Functional Regulatory T Cells Accumulate in Aged Hosts"; Gruver, Hudson, and Sempowski, "Immunosenescence of Ageing."

17. Ron-Harel et al., "Age-Dependent Spatial Memory Loss."

18. Ibid.; Ron-Harel and Schwartz, "Immune Senescence and Brain Aging."

19. Baruch et al., "CNS-Specific Immunity at the Choroid Plexus"; Baruch and Schwartz, "CNS-Specific T Cells Shape Brain Function."

20. Baruch et al., "CNS-Specific Immunity"; Kunis et al., "IFN-Gamma-Dependent Activation of the Brain's Choroid Plexus"; Baruch and Schwartz, "CNS-Specific T Cells Shape Brain Function"; Shechter, London, and Schwartz, "Orchestrated Leukocyte Recruitment to Immune-Privileged Sites."

21. Shechter et al., "Recruitment of Beneficial M2 Macrophages to Injured Spinal Cord."

22. Serot, Bene, and Faure, "Choroid Plexus, Aging of the Brain, and Alzheimer's Disease"; Baruch et al., "CNS-Specific Immunity at the Choroid Plexus"; Baruch and Schwartz, "CNS-Specific T Cells Shape Brain Function"; Cockcroft and Davis, "Mechanisms of Airway Hyperresponsiveness"; Ahdieh, Vandenbos, and Youakim, "Lung Epithelial Barrier Function and Wound Healing"; Coyne et al., "Regulation of Airway Tight Junctions by Proinflammatory Cytokines."

23. Baruch et al., "Aging-Induced Type I Interferon Response."

24. Villeda et al., "The Ageing Systemic Milieu"; Baruch et al., "Aging-Induced Type I Interferon Response."

25. Rogers et al., "Exercise Enhances Vaccine-Induced Antigen-Specific T Cell Responses"; Messaoudi et al., "Delay of T Cell Senescence by Caloric Restriction."

3. STRESS AND DEPRESSION

1. Cannon, *Bodily Changes in Pain, Hunger, Fear, and Rage;* McEwen, "Physiology and Neurobiology of Stress and Adaptation"; Grohol, "What's the Purpose of the Fight or Flight Response?"

2. Sapolsky, *Why Zebras Don't Get Ulcers.*

3. McEwen and Schmeck, *The Hostage Brain;* McEwen, "Physiology and Neurobiology of Stress and Adaptation"; Weissmann, "The Experimental Pathology of Stress."

4. Styron, *Darkness Visible;* Cox, "Illuminating Depression."

5. Lawlor, *From Melancholia to Prozac;* Radden, *The Nature of Melancholy;* Nemade, Reiss, and Dombeck, "Depression"; Mukherjee, "Post-Prozac Nation"; Trede, "150 Years of Freud-Kraepelin Dualism"; Malberg et al., "Chronic Antidepressant Treatment Increases Neurogenesis"; Castren and Hen, "Neuronal Plasticity and Antidepressant Actions."

6. Pappas, "Good Stress Response Enhances Recovery from Surgery"; Dhabhar and McEwen, "Acute Stress Enhances while Chronic Stress Suppresses Cell-Mediated Immunity"; Dhabhar, The Positive Effects of Stress; Dhabhar, "A Hassle a Day May Keep the Pathogens Away."

7. Cohen et al., "Maladaptation to Mental Stress Mitigated by the Adaptive Immune System"; National Institute of Mental Health, "Post-Traumatic Stress Disorder (PTSD)."

8. Cohen et al., "Maladaptation to Mental Stress Mitigated by the Adaptive Immune System"; Lewitus, Cohen, and Schwartz, "Reducing Post-Traumatic Anxiety by Immunization"; Kimpton, "The Brain Derived Neurotrophic Factor"; Lewitus and Schwartz, "Behavioral Immunization."

9. Driscoll, Straus, and Armed Forces Foundation (U.S.), *Hidden Battles on Unseen Fronts;* "PTSD: A Growing Epidemic"; Kawamura, Kim, and Asukai, "Suppression of Cellular Immunity"; Spivak et al., "Elevated Levels of Serum Interleukin-1 Beta"; M. Uddin et al., "Epigenetic and Immune Function Profiles"; Segman et al., "Peripheral Blood Mononuclear Cell Gene Expression Profiles"; Smith et al., "Differential Immune System DNA Methylation and Cytokine Regulation"; Rochelle, "Traumatic Experiences Weaken Immune-System Gene."

10. Silberman et al., "Impaired T-Cell Dependent Humoral Response"; Dhabhar and McEwen, "Acute Stress Enhances while Chronic Stress Suppresses Cell-Mediated Immunity."

11. Saul et al., "Chronic Stress and Susceptibility to Skin Cancer"; Cohen et al., "Chronic Stress."

12. Lewitus et al., "Vaccination as a Novel Approach for Treating Depressive Behavior"; Lewitus, Cohen, and Schwartz, "Reducing Post-Traumatic Anxiety by Immunization"; Smith et al., "Stress and Glucocorticoids"; Malberg et al., "Chronic Antidepressant Treatment Increases Neurogenesis."

13. Silverstein, "Ilya Metchnikoff."

14. Schwartz and Baruch, "Vaccine for the Mind"; Schwartz and Shechter, "Protective Autoimmunity Functions by Intracranial Immunosurveillance"; Lewitus and Schwartz, "Behavioral Immunization."

15. Lewitus, Cohen, and Schwartz, "Reducing Post-Traumatic Anxiety by Immunization."

16. Schwartz and Baruch, "Vaccine for the Mind"; Schwartz and Shechter, "Protective Autoimmunity Functions by Intracranial Immunosurveillance"; Shechter et al., "Recruitment of Beneficial M2 Macrophages to Injured Spinal Cord"; Shechter, London, and Schwartz, "Orchestrated Leukocyte Recruitment to Immune-Privileged Sites."

17. University of Maryland Medical System, "Omega-3 Fatty Acids."

18. Kiecolt-Glaser et al., "Omega-3 Supplementation Lowers Inflammation and Anxiety in Medical Students."

19. Aranow, "Vitamin D and the Immune System."

20. Ziv et al., "Immune Cells Contribute to the Maintenance of Neurogenesis"; Ratey and Hagerman, *Spark*.

21. Glynn, "Exercise Helps Immune System Protect against Future Cancers"; "Exercise and Immunity."

22. "Exercise and Immunity."

4. OF MICE AND SUPERMAN

1. Lee-Liu et al., "Spinal Cord Regeneration."

2. Lee et al., "The Global Map for Traumatic Spinal Cord Injury Epidemiology"; Centers for Disease Control and Prevention, "Spinal Cord Injury."

3. National Institute of Neurological Disorders and Stroke, "Spinal Cord Injury."

4. Hughes, "The Edwin Smith Surgical Papyrus"; Schiller and Mobbs, "The Historical Evolution of the Management of Spinal Cord Injury"; Ramón y Cajal, *Degeneration and Regeneration of the Nervous System.*

5. Kao, "Comparison of Healing Process in Transected Spinal Cords"; Vikhanski, *In Search of the Lost Cord.*

6. Vikhanski, *In Search of the Lost Cord;* David and Aguayo, "Axonal Elongation into Peripheral Nervous System 'Bridges'"; Richardson, McGuinness, and Aguayo, "Axons from CNS Neurons Regenerate into PNS Grafts."

7. Schwab, "Functions of Nogo Proteins and Their Receptors."

8. Rolls, Shechter, and Schwartz, "The Bright Side of the Glial Scar"; Rolls et al., "Two Faces of Chondroitin Sulfate Proteoglycan"; Raposo and Schwartz, "Glial Scar and Immune Cell Involvement"; Shechter et al., "The Glial Scar–Monocyte Interplay."

9. Lee-Liu et al., "Spinal Cord Regeneration"; Schwartz et al., "Regenerating Fish Optic Nerves."

10. Young and Flamm, "Effect of High-Dose Corticosteroid Therapy"; Bracken et al., "A Randomized, Controlled Trial of Methylprednisolone."

11. Hall and Springer, "Neuroprotection and Acute Spinal Cord Injury."

12. Young, "Fear of Hope"; Vikhanski, *In Search of the Lost Cord.*

13. Hall and Springer, "Neuroprotection and Acute Spinal Cord

Injury"; Stevens et al., "Critical Care and Perioperative Management in Traumatic Spinal Cord Injury"; Ito et al., "Does High Dose Methylprednisolone Sodium Succinate Really Improve Neurological Status?"; Short, El Masry, and Jones, "High Dose Methylprednisolone in the Management of Acute Spinal Cord Injury"; Hurlbert et al., "Pharmacological Therapy for Acute Spinal Cord Injury"; Chin et al., "Spinal Cord Injuries Treatment and Management."

14. Crowe et al., "Apoptosis and Delayed Degeneration after Spinal Cord Injury"; Faden, "Experimental Neurobiology of Central Nervous System Trauma."

15. Fitch et al., "Cellular and Molecular Mechanisms of Glial Scarring and Progressive Cavitation"; Popovich et al., "Depletion of Hematogenous Macrophages"; Perry, Andersson, and Gordon, "Macrophages and Inflammation in the Central Nervous System"; Martin, "Wound Healing"; Adamson, "Role of Macrophages in Normal Wound Healing"; Mercandetti, Cohen, and Molnar, "Wound Healing and Repair."

16. Mercandetti, Cohen, and Molnar, "Wound Healing and Repair."

17. Rapalino et al., "Implantation of Stimulated Homologous Macrophages"; Bomstein et al., "Features of Skin-Coincubated Macrophages That Promote Recovery from Spinal Cord Injury."

18. Knoller et al., "Clinical Experience Using Incubated Autologous Macrophages."

19. Ibid.; Proneuron Biotechnologies.

20. Vikhanski, *In Search of the Lost Cord;* "Melissa's Story"; Proneuron Biotechnologies.

21. Knoller et al., "Clinical Experience Using Incubated Autologous Macrophages."

22. Silverstein, "Christopher Reeve"; Weizmann Institute of Science, "Christopher Reeve Comes to Weizmann Institute"; "Barbara

Walters' Last Interview with Christopher Reeve"; "CNN Larry King Live: Interview with Christopher Reeve."

23. "CNN Larry King Live: Interview with Christopher Reeve."

24. Pieribone and Gruber, *Aglow in the Dark;* Shechter et al., "Infiltrating Blood-Derived Macrophages Are Vital Cells"; Jung et al., "Analysis of Fractalkine Receptor CX(3)CR1 Function."

25. Shechter et al., "Infiltrating Blood-Derived Macrophages Are Vital Cells."

26. Raposo and Schwartz, "Glial Scar and Immune Cell Involvement"; Shechter et al., "The Glial Scar–Monocyte Interplay"; Rolls et al., "Two Faces of Chondroitin Sulfate Proteoglycan"; Rolls, Shechter, and Schwartz, "The Bright Side of the Glial Scar"; Fawcett et al., "Defeating Inhibition of Regeneration"; Fawcett, "Overcoming Inhibition in the Damaged Spinal Cord"; Silver and Miller, "Regeneration beyond the Glial Scar."

27. Shechter et al., "Recruitment of Beneficial M2 Macrophages to Injured Spinal Cord"; Shechter, London, and Schwartz, "Orchestrated Leukocyte Recruitment to Immune-Privileged Sites."

28. ImmunoBrain Therapies.

29. Schwartz, "Lecture by Professor Michal Schwartz."

5. A VACCINATION TO PREVENT BLINDNESS

1. Rothstein, "'Dialog in the Dark' at South Street Seaport." See also dialogue-in-the-dark.com.

2. "Visual Impairment and Blindness."

3. Strominger, Demarest, and Laemle, *Noback's Human Nervous System;* Society for Neuroscience, *Brain Facts.*

4. Wurtz, "Retrospective"; Hubel and Wiesel, *Brain and Visual Perception.*

5. Gilbert, *Developmental Biology;* London, Benhar, and Schwartz, "The Retina as a Window to the Brain"; Society for Neuroscience,

Brain Facts; Strominger, Demarest, and Laemle, *Noback's Human Nervous System.*

6. Streilein, "Ocular Immune Privilege"; Shechter, London, and Schwartz, "Orchestrated Leukocyte Recruitment to Immune-Privileged Sites"; London, Benhar, and Schwartz, "The Retina as a Window to the Brain."

7. Faden, "Experimental Neurobiology of Central Nervous System Trauma"; Lynch and Dawson, "Secondary Mechanisms in Neuronal Trauma"; Faden and Salzman, "Pharmacological Strategies in CNS Trauma."

8. Schwartz and Kipnis, "Model of Acute Injury to Study Neuroprotection"; Yoles and Schwartz, "Degeneration of Spared Axons"; Yoles and Schwartz, "Potential Neuroprotective Therapy for Glaucomatous Optic Neuropathy," *Surv Ophthalmol* 42, no. 4 (1998).

9. National Eye Institute, "Facts about Glaucoma."

10. Brubaker, "Delayed Functional Loss in Glaucoma"; Werner and Drance, "Progression of Glaucomatous Field Defects despite Successful Filtration"; Freudenthal et al., "Low-Tension Glaucoma."

11. Schwartz et al., "Potential Treatment Modalities for Glaucomatous Neuropathy"; Dreyer et al., "Elevated Glutamate Levels in the Vitreous Body of Humans and Monkeys with Glaucoma"; Schwartz and Yoles, "Neuroprotection"; Yoles and Schwartz, "Potential Neuroprotective Therapy for Glaucomatous Optic Neuropathy"; Yoles and Schwartz, "Degeneration of Spared Axons"; Schwartz and Yoles, "Optic Nerve Degeneration and Potential Neuroprotection."

12. Levin and Peeples, "History of Neuroprotection and Rationale as a Therapy for Glaucoma"; Weinreb and Levin, "Is Neuroprotection a Viable Therapy for Glaucoma?"

13. Yoles et al., "Protective Autoimmunity Is a Physiological Response to CNS Trauma"; Moalem et al., "Autoimmune T Cells Protect Neurons from Secondary Degeneration"; Schori, Yoles, and

Schwartz, "T-Cell-Based Immunity Counteracts the Potential Toxicity of Glutamate."

14. Kipnis et al., "T Cell Immunity to Copolymer 1 Confers Neuroprotection"; Schwartz, "Physiological Approaches to Neuro-protection"; Schwartz, "Harnessing the Immune System for Neuro-protection"; Schwartz et al., "Protective Autoimmunity against the Enemy Within"; Ben Simon et al., "A Rat Model for Acute Rise in Intraocular Pressure."

15. Barouch and Schwartz, "Autoreactive T Cells Induce Neuro-trophin Production"; Shechter et al., "Infiltrating Blood-Derived Macrophages Are Vital Cells"; London et al., "Neuroprotection and Progenitor Cell Renewal."

16. Schwartz, "Neurodegeneration and Neuroprotection in Glaucoma."

17. Schwartz, "Physiological Approaches to Neuroprotection"; Kipnis et al., "T Cell Immunity to Copolymer 1 Confers Neuro-protection"; Schwartz, "Harnessing the Immune System for Neuro-protection"; Kipnis and Schwartz, "Dual Action of Glatiramer Acetate"; Schwartz, "Vaccination for Glaucoma."

18. Mitne et al., "The Potential Neuroprotective Effects of Weekly Treatment with Glatiramer Acetate."

19. London, Benhar, and Schwartz, "The Retina as a Window to the Brain."

6. ALZHEIMER'S DISEASE AND LOU GEHRIG'S DISEASE (ALS)

1. American Cancer Society, "Rising Global Cancer Epidemic."

2. Zea et al., "Arginase-Producing Myeloid Suppressor Cells"; Mantovani et al., "Cancer-Related Inflammation"; Grivennikov, Greten, and Karin, "Immunity, Inflammation, and Cancer."

3. Schwartz and Ziv, "Immunity to Self and Self-Maintenance."

4. Harvard School of Public Health, "International Survey."

5. Mace and Rabins, *The 36-Hour Day*.

6. Cohen and Eisdorfer, *The Loss of Self*.

7. National Institute of Neurological Disorders and Stroke, "Amyotrophic Lateral Sclerosis (ALS) Fact Sheet"; Lou Gehrig Web site.

8. National Institute on Aging, "Alzheimer's Disease."

9. National Institute on Aging, "Alzheimer's Disease Medications Fact Sheet"; Berk and Sabbagh, "Successes and Failures."

10. Renton, Chio, and Traynor, "State of Play in Amyotrophic Lateral Sclerosis Genetics"; Rosen et al., "Mutations in Cu/Zn Superoxide Dismutase Gene"; Ripps et al., "Transgenic Mice Expressing an Altered Murine Superoxide Dismutase Gene"; Renton et al., "A Hexanucleotide Repeat Expansion in C9ORF72"; DeJesus-Hernandez et al., "Expanded GGGGCC Hexanucleotide Repeat"; Donnelly et al., "RNA Toxicity from the ALS/FTD C9ORF72 Expansion."

11. Armon and Lorenzo, "Amyotrophic Lateral Sclerosis"; Gibson and Bromberg, "Amyotrophic Lateral Sclerosis."

12. Armon and Lorenzo, "Amyotrophic Lateral Sclerosis."

13. Weisskopf et al., "Prospective Study of Military Service and Mortality from ALS"; Armon and Lorenzo, "Amyotrophic Lateral Sclerosis"; "VA Secretary Establishes ALS as a Presumptive Compensable Illness."

14. Hardiman, van den Berg, and Kiernan, "Clinical Diagnosis and Management of Amyotrophic Lateral Sclerosis"; Gibson and Bromberg, "Amyotrophic Lateral Sclerosis."

15. Gao and Hong, "Why Neurodegenerative Diseases Are Progressive."

16. Masters, Wyss-Coray, and Pasinetti, "Anti-Inflammatory Drugs Fall Short in Alzheimer's Disease"; Gordon et al., "Efficacy of Minocycline in Patients with Amyotrophic Lateral Sclerosis"; Schwartz and Shechter, "Systemic Inflammatory Cells Fight Off Neurodegenerative Disease."

17. Schwartz and Shechter, "Systemic Inflammatory Cells Fight off Neurodegenerative Disease"; Schwartz and Baruch, "The Resolution of Neuroinflammation in Neurodegeneration."

18. Aisen and Davis, "Inflammatory Mechanisms in Alzheimer's Disease."

19. Simard et al., "Bone Marrow–Derived Microglia Play a Critical Role."

20. Butovsky et al., "Glatiramer Acetate Fights against Alzheimer's Disease"; Koronyo-Hamaoui et al., "Attenuation of AD-Like Neuropathology by Harnessing Peripheral Immune Cells"; Butovsky et al., "Selective Ablation of Bone Marrow–Derived Dendritic Cells."

21. Shechter et al., "Recruitment of Beneficial M2 Macrophages to Injured Spinal Cord"; K. Baruch, N. Rosenzweig, et al., unpublished results.

22. Baruch, Rosenzweig, et al., unpublished results; Almand et al., "Increased Production of Immature Myeloid Cells"; Zea et al., "Arginase-Producing Myeloid Suppressor Cells"; Mantovani et al., "Cancer-Related Inflammation."

23. Shechter et al., "Infiltrating Blood-Derived Macrophages Are Vital Cells"; Baruch, Rosenzweig, et al., unpublished results.

24. Butovsky et al., "Glatiramer Acetate Fights against Alzheimer's Disease"; Butovsky et al., "Selective Ablation of Bone Marrow–Derived Dendritic Cells"; Baruch, Rosenzweig, et al., unpublished results.

25. Baruch, Rosenzweig, et al., unpublished results; Koronyo-Hamaoui et al., "Attenuation of AD-Like Neuropathology by Harnessing Peripheral Immune Cells"; Butovsky et al., "Selective Ablation of Bone Marrow–Derived Dendritic Cells"; Butovsky et al., "Glatiramer Acetate Fights against Alzheimer's Disease."

26. Beers et al., "CD4+ T Cells Support Glial Neuroprotection"; Banerjee et al., "Adaptive Immune Neuroprotection."

27. Vaknin et al., "Excess Circulating Alternatively Activated Myeloid (M2) Cells"; Seksenyan et al., "Thymic Involution."

28. G. Kunis, et al., "Immunization with a Myelin-Derived Antigen Activates the Brain's Choroid Plexus."

29. Baruch and Schwartz, "CNS-Specific T Cells Shape Brain Function"; Schwartz and Baruch, "The Resolution of Neuroinflammation in Neurodegeneration."

30. Schwartz and Shechter, "Systemic Inflammatory Cells Fight Off Neurodegenerative Disease."

31. Kunis, Baruch, et al., unpublished results.

32. Orgogozo et al., "Subacute Meningoencephalitis in a Subset of Patients with AD after Abeta42 Immunization"; Winblad et al., "Safety, Tolerability, and Antibody Response."

33. Stiles and Jernigan, "The Basics of Brain Development"; Schwarz and Bilbo, *The Immune System and the Developing Brain;* Bilbo and Schwarz, "The Immune System and Developmental Programming."

34. Velasquez-Manoff, "An Immune Disorder at the Root of Autism"; Krause and Muller, "The Relationship between Tourette's Syndrome and Infections."

35. Derecki et al., "Wild-Type Microglia Arrest Pathology."

36. Cardon et al., "Dysregulation of Kisspeptin and Neurogenesis at Adolescence."

7. MALES' AND FEMALES' DIFFERENT IMMUNE SYSTEMS

1. Summers, "Lawrence (Larry) Summers on Women in Science"; Valian, "Raise Your Hand if You're a Woman in Science"; Bombardieri, "Summers' Remarks on Women Draw Fire"; Summers, "Remarks at NBER Conference."

2. Carter et al., "Sex and Gender Differences in Alzheimer's Disease"; Miller and Cronin-Golomb, "Gender Differences in Par-

kinson's Disease"; McCombe and Henderson, "Effects of Gender in Amyotrophic Lateral Sclerosis"; McCarthy et al., "Sex Differences in the Brain."

3. McCarthy et al., "Sex Differences in the Brain."

4. Achiron, Lipitz, and Achiron, "Sex-Related Differences in the Development of the Human Fetal Corpus Callosum"; Bergland, "Scientists Identify Why Girls Often Mature Faster Than Boys"; Lim et al., "Preferential Detachment during Human Brain Development"; "Men and Women Use Brain Differently"; Gurian and Stevens, "Educational Leadership"; "How Male and Female Brains Differ."

5. "Having a Male Co-Twin Improves Mental Rotation Performance in Females."

6. Barkow, Cosmides, and Tooby, *The Adapted Mind;* Eliot, *Pink Brain, Blue Brain.*

7. White, "Girls' Immune Systems Rule"; Oertelt-Prigione, "The Influence of Sex and Gender on the Immune Response"; Furman et al., "Systems Analysis of Sex Differences"; Tarlach, "Why Women's Immune Systems Are Stronger Than Men's."

8. De Groot et al., "Differences in Cognitive Performance during Pregnancy and Early Motherhood"; Buckwalter et al., "Pregnancy and Post Partum"; Christensen, Leach, and Mackinnon, "Cognition in Pregnancy and Motherhood"; Jarrett, "The Truth about the Effect of Pregnancy on Women's Brains."

9. Billington, "The Immunological Problem of Pregnancy"; Luppi, "How Immune Mechanisms Are Affected by Pregnancy."

10. Rolls et al., "Decrease in Hippocampal Neurogenesis during Pregnancy."

11. Hauben et al., "Sexual Dimorphism in the Spontaneous Recovery from Spinal Cord Injury."

12. Oertelt-Prigione, "The Influence of Sex and Gender on the Immune Response"; Furman et al., "Systems Analysis of Sex Dif-

ferences"; Hauben et al., "Sexual Dimorphism in the Spontaneous Recovery from Spinal Cord Injury."

13. Stein, "Brain Damage, Sex Hormones, and Recovery."

14. Whitacre, "Sex Differences in Autoimmune Disease"; Fairweather, Frisancho-Kiss, and Rose, "Sex Differences in Autoimmune Disease"; Oliver and Silman, "Why Are Women Predisposed to Autoimmune Rheumatic Diseases?"; Carter et al., "Sex and Gender Differences in Alzheimer's Disease."

15. McCombe and Henderson, "Effects of Gender in Amyotrophic Lateral Sclerosis."

16. McCarthy et al., "Sex Differences in the Brain."

A NEUROIMMUNOLOGY PRIMER

1. Strominger, Demarest, and Laemle, *Noback's Human Nervous System;* Society for Neuroscience, *Brain Facts;* Newquist, Kasnot, and Brace, *The Great Brain Book;* Hsu, "How Much Power?"

2. Society for Neuroscience, *Brain Facts;* Strominger, Demarest, and Laemle, *Noback's Human Nervous System;* National Institute of Mental Health, National Institutes of Health, U.S. Department of Health and Human Services, "Brain Basics."

3. Society for Neuroscience, *Brain Facts;* Rappoport, "Your Mind Is Not a Computer."

4. Society for Neuroscience, *Brain Facts.*

5. Ibid.; Strominger, Demarest, and Laemle, *Noback's Human Nervous System;* London, Benhar, and Schwartz, "The Retina as a Window to the Brain."

6. Shechter, London, and Schwartz, "Orchestrated Leukocyte Recruitment to Immune-Privileged Sites"; McCall, Millington, and Wurtman, "Blood-Brain Barrier Transport of Caffeine"; Chen, Ghribi, and Geiger, "Caffeine Protects against Disruptions of the Blood-Brain

Barrier"; Haorah et al., "Alcohol-Induced Oxidative Stress in Brain Endothelial Cells"; Ransohoff and Engelhardt, "The Anatomical and Cellular Basis of Immune Surveillance."

7. Strominger, Demarest, and Laemle, *Noback's Human Nervous System;* Society for Neuroscience, *Brain Facts;* Shechter, London, and Schwartz, "Orchestrated Leukocyte Recruitment to Immune-Privileged Sites."

8. Shechter, London, and Schwartz, "Orchestrated Leukocyte Recruitment to Immune-Privileged Sites"; Bernd, "Epithelial Cells Introduction"; Strominger, Demarest, and Laemle, *Noback's Human Nervous System.*

9. Sompayrac, *How the Immune System Works.*

10. Ibid.; Shechter and Schwartz, "Harnessing Monocyte-Derived Macrophages to Control Central Nervous System Pathologies."

11. Sompayrac, *How the Immune System Works;* Gowans, McGregor, and Cowen, "Initiation of Immune Responses by Small Lymphocytes"; Greenberg, "A Concise History of Immunology."

12. Sompayrac, *How the Immune System Works.*

13. Rosenberg et al., "Immunologic and Therapeutic Evaluation of a Synthetic Peptide Vaccine"; Wagstaff, "Therapeutic Cancer Vaccines."

14. Greenberg, "A Concise History of Immunology."

15. Moalem et al., "Autoimmune T Cells Protect Neurons from Secondary Degeneration"; Schwartz and Raposo, "Protective Autoimmunity"; Schwartz and Shechter, "Protective Autoimmunity Functions by Intracranial Immunosurveillance to Support the Mind"; Schwartz and Ziv, "Immunity to Self and Self-Maintenance"; Schwartz and Kipnis, "Protective Autoimmunity and Neuroprotection"; "Protective Autoimmunity."

16. Sompayrac, *How the Immune System Works;* Schwartz and Ziv, "Immunity to Self and Self-Maintenance."

17. Tauber, "Metchnikoff and the Phagocytosis Theory."

18. Zipp and Aktas, "The Brain as a Target of Inflammation"; Masters, Wyss-Coray, and Pasinetti, "Anti-Inflammatory Drugs Fall Short in Alzheimer's Disease."

BIBLIOGRAPHY

ACHIRON, R., S. LIPITZ, and A. ACHIRON. "Sex-Related Differences in the Development of the Human Fetal Corpus Callosum: In Utero Ultrasonographic Study." *Prenatal Diagnosis* 21, no. 2 (2001): 116–20.

ADAMSON, R. "Role of Macrophages in Normal Wound Healing: An Overview." *Journal of Wound Care* 18, no. 8 (2009): 349–51.

AHDIEH, M., T. VANDENBOS, and A. YOUAKIM. "Lung Epithelial Barrier Function and Wound Healing Are Decreased by IL-4 and IL-13 and Enhanced by IFN-Gamma." *American Journal of Physiology. Cell Physiology* 281, no. 6 (2001): C2029–38.

AISEN, P. S., and K. L. DAVIS. "Inflammatory Mechanisms in Alzheimer's Disease: Implications for Therapy." *The American Journal of Psychiatry* 151, no. 8 (1994): 1105–13.

ALMAND, B., J. I. CLARK, E. NIKITINA, J. VAN BEYNEN, N. R. ENGLISH, S. C. KNIGHT, D. P. CARBONE, and D. I. GABRILOVICH. "Increased Production of Immature Myeloid Cells in Cancer Patients: A Mechanism of Immunosuppression in Cancer." *Journal of Immunology* 166, no. 1 (2001): 678–89.

ALTMAN, J. "Are New Neurons Formed in the Brains of Adult Mammals?" *Science* 135, no. 3509 (1962): 1127–28.

AMERICAN CANCER SOCIETY. "Rising Global Cancer Epidemic." http://www.cancer.org/research/infographicgallery/rising-global-cancer-epidemic-text-alternative.

ARANOW, C. "Vitamin D and the Immune System." *Journal of Investigative Medicine* 59, no. 6 (2011): 881–86.

ARMON, CARMEL, and NICHOLAS LORENZO. "Amyotrophic Lateral Sclerosis." *Medscape* (2014), http://emedicine.medscape.com/article/1170097-overview#aw2aab6b2b3.

BANERJEE, R., R. L. MOSLEY, A. D. REYNOLDS, A. DHAR, V. JACKSON-LEWIS, P. H. GORDON, S. PRZEDBORSKI, and H. E. GENDELMAN. "Adaptive Immune Neuroprotection in G93A-SOD1 Amyotrophic Lateral Sclerosis Mice." *PLoS One* 3, no. 7 (2008): e2740.

"BARBARA WALTERS' LAST INTERVIEW WITH CHRISTOPHER REEVE." ABC News (2004), http://abcnews.go.com/2020/ABCNEWSSpecial/story?id=124364&page=1.

BARKOW, JEROME H., LEDA COSMIDES, and JOHN TOOBY. *The

Adapted Mind: Evolutionary Psychology and the Generation of Culture. New York: Oxford University Press, 1992.

BAROUCH, R., and M. SCHWARTZ. "Autoreactive T Cells Induce Neurotrophin Production by Immune and Neural Cells in Injured Rat Optic Nerve: Implications for Protective Autoimmunity." *FASEB Journal* 16, no. 10 (2002): 1304–6.

BARUCH, K., A. DECZKOWSKA, E. DAVID, J. M. CASTELLANO, O. MILLER, A. KERTSER, T. BERKUTZKI, Z. BARNETT-ITZHAKI, D. BEZALEL, T. WYSS-CORAY, I. AMIT, and M. SCHWARTZ. "Aging-Induced Type I Interferon Response at the Choroid Plexus Negatively Affects Brain Function." *Science* 346, no. 6205 (2014): 89–93.

BARUCH, K., N. RON-HAREL, H. GAL, A. DECZKOWSKA, E. SHIFRUT, W. NDIFON, N. MIRLAS-NEISBERG, M. CARDON, I. VAKNIN, L. CAHALON, T. BERKUTZKI, M. P. MATTSON, F. GOMEZ-PINILLA, N. FRIEDMAN, and M. SCHWARTZ. "CNS-Specific Immunity at the Choroid Plexus Shifts toward Destructive Th2 Inflammation in Brain Aging." *Proceedings of the National Academy of Sciences of the USA* 110, no. 6 (2013): 2264–69.

BARUCH, K., and M. SCHWARTZ. "CNS-Specific T Cells Shape Brain Function via the Choroid Plexus." *Brain, Behavior, and Immunity* 34 (2013): 11–16.

BEERS, D. R., J. S. HENKEL, W. ZHAO, J. WANG, and S. H. APPEL. "CD4+ T Cells Support Glial Neuroprotection, Slow Disease Progression, and Modify Glial Morphology in an Animal Model of Inherited ALS." *Proceedings of the National Academy of Sciences of the USA* 105, no. 40 (2008): 15558–63.

BEN SIMON, G. J., S. BAKALASH, E. ALONI, and M. ROSNER. "A Rat Model for Acute Rise in Intraocular Pressure: Immune Modulation as a Therapeutic Strategy." *American Journal of Ophthalmology* 141, no. 6 (2006): 1105–11.

BERAL, VALERIE, EMILY BANKS, GILLIAN REEVES, and DIANA BULL. "Breast Cancer and Hormone-Replacement Therapy: The Million Women Study." *Lancet* 362 (2003): 1330–31.

BERGLAND, CHRISTOPHER. "Scientists Identify Why Girls Often Mature Faster Than Boys." *Psychology Today* (2013), http://www.psychologytoday.com/blog/the-athletes -way/201312/scientists-identify-why-girls-often-mature -faster-boys.

BERK, C., and M. N. SABBAGH. "Successes and Failures for Drugs in Late-Stage Development for Alzheimer's Disease." *Drugs Aging* 30, no. 10 (2013): 783–92.

BERNARDES DE JESUS, B., E. VERA, K. SCHNEEBERGER, A. M. TEJERA, E. AYUSO, F. BOSCH, and M. A. BLASCO. "Telomerase Gene Therapy in Adult and Old Mice Delays Aging and In-creases Longevity without Increasing Cancer." *EMBO Molecular Medicine* 4, no. 8 (2012): 691–704.

BERND, KAREN. "Epithelial Cells Introduction." Davidson College Biology Department, 2010, http://www.bio .davidson.edu/people/kabernd/berndcv/lab/epithelial infoweb/index.html.

BILBO, S. D., and J. M. SCHWARZ. "The Immune System and Developmental Programming of Brain and Behavior." *Frontiers in Neuroendocrinology* 33, no. 3 (2012): 267–86.

BILLINGTON, W. D. "The Immunological Problem of Pregnancy: 50 Years with the Hope of Progress. A Tribute to Peter Medawar." *Journal of Reproductive Immunology* 60, no. 1 (2003): 1–11.

BOMBARDIERI, MARCELLA. "Summers' Remarks on Women Draw Fire." *Boston Globe,* January 17, 2005, http://www .boston.com/news/education/higher/articles/2005/ 01/17/summers_remarks_on_women_draw_fire/ ?page=2.

BOMSTEIN, Y., J. B. MARDER, K. VITNER, I. SMIRNOV, G. LISAEY, O. BUTOVSKY, V. FULGA, and E. YOLES. "Features of Skin-Coincubated Macrophages That Promote Recovery from Spinal Cord Injury." *Journal of Neuroimmunology* 142, nos. 1–2 (2003): 10–16.

BRACKEN, M. B., M. J. SHEPARD, W. F. COLLINS, T. R. HOLFORD, W. YOUNG, D. S. BASKIN, H. M. EISENBERG, E. FLAMM, L. LEO-SUMMERS, J. MAROON, et al. "A Randomized, Controlled Trial of Methylprednisolone or Naloxone in the Treatment of Acute Spinal-Cord Injury. Results of the Second National Acute Spinal Cord Injury Study." *New England Journal of Medicine* 322, no. 20 (1990): 1405–11.

BRUBAKER, R. F. "Delayed Functional Loss in Glaucoma. LII Edward Jackson Memorial Lecture." *American Journal of Ophthalmology* 121, no. 5 (1996): 473–83.

BUCKWALTER, J. G., D. K. BUCKWALTER, B. W. BLUESTEIN, and F. Z. STAN-CZYK. "Pregnancy and Post Partum: Changes in Cognition and Mood." *Progress in Brain Research* 133 (2001): 303–19.

BUTOVSKY, O., M. KORONYO-HAMAOUI, G. KUNIS, E. OPHIR, G. LANDA, H. COHEN, and M. SCHWARTZ. "Glatiramer Acetate Fights against Alzheimer's Disease by Inducing Dendritic-Like Microglia Expressing Insulin-Like Growth Factor 1." *Proceedings of the National Academy of Sciences of the USA* 103, no. 31 (2006): 11784–89.

BUTOVSKY, O., G. KUNIS, M. KORONYO-HAMAOUI, and M. SCHWARTZ. "Selective Ablation of Bone Marrow-Derived Dendritic Cells Increases Amyloid Plaques in a Mouse Alzheimer's Disease Model." *The European Journal of Neuroscience* 26, no. 2 (2007): 413–16.

CAHN, B. R., and J. POLICH. "Meditation States and Traits: EEG, ERP, and Neuroimaging Studies." *Psychological Bulletin* 132, no. 2 (2006): 180–211.

CANNON, WALTER B. *Bodily Changes in Pain, Hunger, Fear, and Rage: An Account of Recent Researches into the Function of Emotional Excitement*. New York: Appleton, 1915.

CARDON, M., N. RON-HAREL, H. COHEN, G. M. LEWITUS, and M. SCHWARTZ. "Dysregulation of Kisspeptin and Neurogenesis at Adolescence Link Inborn Immune Deficits to the Late Onset of Abnormal Sensorimotor Gating in Congenital Psychological Disorders." *Molecular Psychiatry* 15, no. 4 (2010): 415–25.

CARTER, C. L., E. M. RESNICK, M. MALLAMPALLI, and A. KALBARCZYK. "Sex and Gender Differences in Alzheimer's Disease: Recommendations for Future Research." *Journal of Women's Health* 21, no. 10 (2012): 1018–23.

CASTREN, E., and R. HEN. "Neuronal Plasticity and Antidepressant Actions." *Trends in Neurosciences* 36, no. 5 (2013): 259–67.

CENTERS FOR DISEASE CONTROL AND PREVENTION. "Spinal Cord Injury (Sci): Fact Sheet." http://www.cdc.gov/traumatic braininjury/scifacts.html.

CHEN, X., O. GHRIBI, and J. D. GEIGER. "Caffeine Protects against Disruptions of the Blood-Brain Barrier in Animal Models of Alzheimer's and Parkinson's Diseases." *Journal of Alzheimer's Disease* 20, suppl. 1 (2010): S127–41.

CHIN, LAWRENCE S., FASSIL B. MESFIN, SEGUN T. DAWODU, and ALLEN R. WYLER. "Spinal Cord Injuries Treatment and Management." *Medscape* (2014), http://emedicine.med scape.com/article/793582-treatment#aw2aab6b6b4.

CHRISTENSEN, H., L. S. LEACH, and A. MACKINNON. "Cognition in Pregnancy and Motherhood: Prospective Cohort Study." *British Journal of Psychiatry* 196, no. 2 (2010): 126–32.

CLARKE, DESMOND M. *Descartes: A Biography*. New York: Cambridge University Press, 2006.

"CNN LARRY KING LIVE: INTERVIEW WITH CHRISTOPHER REEVE." CNN.com (2003), http://transcripts.cnn.com/TRAN SCRIPTS/0307/30/lkl.00.html.

COCKCROFT, D. W., and B. E. DAVIS. "Mechanisms of Airway Hyperresponsiveness." *Journal of Allergy and Clinical Immunology* 118, no. 3 (2006): 551–59, quiz 560–61.

COHEN, DONNA, and CARL EISDORFER. *The Loss of Self: A Family Resource for the Care of Alzheimer's Disease and Related Disorders*. Rev. and updated ed. New York: Norton, 2001.

COHEN, H., Y. ZIV, M. CARDON, Z. KAPLAN, M. A. MATAR, Y. GIDRON, M. SCHWARTZ, and J. KIPNIS. "Maladaptation to Mental Stress Mitigated by the Adaptive Immune System via Depletion of Naturally Occurring Regulatory CD4+CD25+ Cells." *Journal of Neurobiology* 66, no. 6 (2006): 552–63.

COHEN, S., D. JANICKI-DEVERTS, W. J. DOYLE, G. E. MILLER, E. FRANK, B. S. RABIN, and R. B. TURNER. "Chronic Stress, Glucocorticoid Receptor Resistance, Inflammation, and Disease Risk." *Proceedings of the National Academy of Sciences of the USA* 109, no. 16 (2012): 5995–99.

CORLISS, RICHARD, and MICHAEL D. LEMONICK. "How to Live to Be 100." *Time*, August 30, 2004, http://content.time.com/time/magazine/article/0,9171,994967,00.html.

COX, CHRIS. "Illuminating Depression." *Guardian*, March 7, 2011, http://www.theguardian.com/books/booksblog/2011/mar/07/illuminating-depression-william-styron.

COYNE, C. B., M. K. VANHOOK, T. M. GAMBLING, J. L. CARSON, R. C. BOUCHER, and L. G. JOHNSON. "Regulation of Airway Tight Junctions by Proinflammatory Cytokines." *Molecular Biology of the Cell* 13, no. 9 (2002): 3218–34.

CROWE, M. J., J. C. BRESNAHAN, S. L. SHUMAN, J. N. MASTERS, and M. S. BEATTIE. "Apoptosis and Delayed Degeneration after Spinal Cord Injury in Rats and Monkeys." *Nature Medicine* 3, no. 1 (1997): 73–76.

DAVID, S., and A. J. AGUAYO. "Axonal Elongation into Peripheral Nervous System 'Bridges' after Central Nervous System Injury in Adult Rats." *Science* 214, no. 4523 (1981): 931–33.

DAVIDSON, R. J., J. KABAT-ZINN, J. SCHUMACHER, M. ROSENKRANZ, D. MULLER, S. F. SANTORELLI, F. URBANOWSKI, A. HARRINGTON, K. BONUS, and J. F. SHERIDAN. "Alterations in Brain and Immune Function Produced by Mindfulness Meditation." *Psychosomatic Medicine* 65, no. 4 (2003): 564–70.

DE GROOT, R. H., E. F. VUURMAN, G. HORNSTRA, and J. JOLLES. "Differences in Cognitive Performance during Pregnancy and Early Motherhood." *Psychological Medicine* 36, no. 7 (2006): 1023–32.

DEJESUS-HERNANDEZ, M., I. R. MACKENZIE, B. F. BOEVE, A. L. BOXER, M. BAKER, N. J. RUTHERFORD, A. M. NICHOLSON, N. A. FINCH, H. FLYNN, J. ADAMSON, N. KOURI, A. WOJTAS, P. SENGDY, G. Y. HSIUNG, A. KARYDAS, W. W. SEELEY, K. A. JOSEPHS, G. COPPOLA, D. H. GESCHWIND, Z. K. WSZOLEK, H. FELDMAN, D. S. KNOPMAN, R. C. PETERSEN, B. L. MILLER, D. W. DICKSON, K. B. BOYLAN, N. R. GRAFF-RADFORD, and R. RADEMAKERS. "Expanded GGGGCC Hexanucleotide Repeat in Noncoding Region of C9ORF72 Causes Chromosome 9p-Linked FTD and ALS." *Neuron* 72, no. 2 (2011): 245–56.

DERECKI, N. C., J. C. CRONK, Z. LU, E. XU, S. B. ABBOTT, P. G. GUYENET, and J. KIPNIS. "Wild-Type Microglia Arrest Pathology in a Mouse Model of Rett Syndrome." *Nature* 484, no. 7392 (2012): 105–9.

DESCARTES, RENÉ, and JOHN COTTINGHAM. *Meditations on First Philosophy: With Selections from the Objections and Replies; A*

Latin-English Edition. Cambridge: Cambridge University Press, 2013.

DHABHAR, FIRDAUS. The Positive Effects of Stress. TED talks (2013), https://www.youtube.com/watch?v=nsc83N -Q1q4.

———. "A Hassle a Day May Keep the Pathogens Away: The Fight-or-Flight Stress Response and the Augmentation of Immune Function." *Integrative and Comparative Biology* 49, no. 3 (2009): 215–36.

DHABHAR, F. S., and B. S. MCEWEN. "Acute Stress Enhances while Chronic Stress Suppresses Cell-Mediated Immunity in Vivo: A Potential Role for Leukocyte Trafficking." *Brain, Behavior, and Immunity* 11, no. 4 (1997): 286–306.

DONNELLY, C. J., P. W. ZHANG, J. T. PHAM, A. R. HAEUSLER, N. A. MISTRY, S. VIDENSKY, E. L. DALEY, E. M. POTH, B. HOOVER, D. M. FINES, N. MARAGAKIS, P. J. TIENARI, L. PETRUCELLI, B. J. TRAYNOR, J. WANG, F. RIGO, C. F. BENNETT, S. BLACKSHAW, R. SATTLER, and J. D. ROTHSTEIN. "RNA Toxicity from the ALS/FTD C9ORF72 Expansion Is Mitigated by Antisense Intervention." *Neuron* 80, no. 2 (2013): 415–28.

DREYER, E. B., D. ZURAKOWSKI, R. A. SCHUMER, S. M. PODOS, and S. A. LIPTON. "Elevated Glutamate Levels in the Vitreous Body of Humans and Monkeys with Glaucoma." *Archives of Ophthalmology* 114, no. 3 (1996): 299–305.

DRISCOLL, PATRICIA P., CELIA STRAUS, and ARMED FORCES FOUNDATION (U.S.). *Hidden Battles on Unseen Fronts: Stories of Amer-*

ican *Soldiers with Traumatic Brain Injury and PTSD.* Drexel Hill, Pa.: Casemate, 2009.

ELIOT, LISE. *Pink Brain, Blue Brain: How Small Differences Grow into Troublesome Gaps—andWhatWe Can Do about It.* Boston: Houghton Mifflin Harcourt, 2009.

ERIKSSON, P. S., E. PERFILIEVA, T. BJORK-ERIKSSON, A. M. ALBORN, C. NORDBORG, D. A. PETERSON, and F. H. GAGE. "Neurogenesis in the Adult Human Hippocampus." *Nature Medicine* 4, no. 11 (1998): 1313–17.

"EXERCISE AND IMMUNITY." MedlinePlus, May 15, 2012, http://www.nlm.nih.gov/medlineplus/ency/article/007165.htm.

FADEN, A. I. "Experimental Neurobiology of Central Nervous System Trauma." *Critical Reviews in Neurobiology* 7, nos. 3–4 (1993): 175–86.

FADEN, A. I., and S. SALZMAN. "Pharmacological Strategies in CNS Trauma." *Trends in Pharmacological Sciences* 13, no. 1 (1992): 29–35.

FAIRWEATHER, D., S. FRISANCHO-KISS, and N. R. ROSE. "Sex Differences in Autoimmune Disease from a Pathological Perspective." *American Journal of Pathology* 173, no. 3 (2008): 600–609.

FAWCETT, J. W. "Overcoming Inhibition in the Damaged Spinal Cord." *Journal of Neurotrauma* 23, nos. 3–4 (2006): 371–83.

FAWCETT, J. W., M. E. SCHWAB, L. MONTANI, N. BRAZDA, and H. W. MULLER. "Defeating Inhibition of Regeneration by

Scar and Myelin Components." *Handbook of Clinical Neurology* 109 (2012): 503–22.

FITCH, M. T., C. DOLLER, C. K. COMBS, G. E. LANDRETH, and J. SILVER. "Cellular and Molecular Mechanisms of Glial Scarring and Progressive Cavitation: In Vivo and in Vitro Analysis of Inflammation-Induced Secondary Injury after CNS Trauma." *Journal of Neuroscience* 19, no. 19 (1999): 8182–98.

FONTANA, L., T. E. MEYER, S. KLEIN, and J. O. HOLLOSZY. "Long-Term Calorie Restriction Is Highly Effective in Reducing the Risk for Atherosclerosis in Humans." *Proceedings of the National Academy of Sciences of the USA* 101, no. 17 (2004): 6659–63.

FREUDENTHAL, JACQUELINE, IQBAL IKE K. AHMED, BASEER U. KHAN, KHALID HASANEE, and HAMPTON ROY SR. "Low-Tension Glaucoma." *Medscape* (2012), http://emedicine.medscape.com/article/1205508-overview.

FURMAN, D., B. P. HEJBLUM, N. SIMON, V. JOJIC, C. L. DEKKER, R. THIEBAUT, R. J. TIBSHIRANI, and M. M. DAVIS. "Systems Analysis of Sex Differences Reveals an Immunosuppressive Role for Testosterone in the Response to Influenza Vaccination." *Proceedings of the National Academy of Sciences of the USA* 111, no. 2 (2014): 869–74.

GAO, H. M., and J. S. HONG. "Why Neurodegenerative Diseases Are Progressive: Uncontrolled Inflammation Drives Disease Progression." *Trends in Immunology* 29, no. 8 (2008): 357–65.

GEDA, Y. E., R. O. ROBERTS, D. S. KNOPMAN, T. J. CHRISTIANSON, V. S. PANKRATZ, R. J. IVNIK, B. F. BOEVE, E. G. TANGALOS, R. C. PETERSEN, and W. A. ROCCA. "Physical Exercise, Aging, and Mild Cognitive Impairment: A Population-Based Study." *Archives of Neurology* 67, no. 1 (2010): 80–86.

GIBSON, S. B., and M. B. BROMBERG. "Amyotrophic Lateral Sclerosis: Drug Therapy from the Bench to the Bedside." *Seminars in Neurology* 32, no. 3 (2012): 173–78.

GILBERT, SCOTT F. *Developmental Biology*. 9th ed. Sunderland, Mass.: Sinauer, 2010.

GLIMCHER, PAUL W. "René Descartes and the Birth of Neuro-science," in *Decisions, Uncertainty, and the Brain: The Science of Neuroeconomics*. Cambridge: MIT Press, 2003.

GLYNN, SARAH. "Exercise Helps Immune System Protect against Future Cancers." *Medical News Today* (2012), http://www.medicalnewstoday.com/articles/251415.php.

GORDON, P. H., D. H. MOORE, R. G. MILLER, J. M. FLORENCE, J. L. VER-HEIJDE, C. DOORISH, J. F. HILTON, G. M. SPITALNY, R. B. MACAR-THUR, H. MITSUMOTO, H. E. NEVILLE, K. BOYLAN, T. MOZAFFAR, J. M. BELSH, J. RAVITS, R. S. BEDLACK, M. C. GRAVES, L. F. MC-CLUSKEY, R. J. BAROHN, R. TANDAN, and A.L.S. STUDY GROUP WESTERN. "Efficacy of Minocycline in Patients with Amyo-trophic Lateral Sclerosis: A Phase III Randomised Trial." *Lancet. Neurology* 6, no. 12 (2007): 1045–53.

GOULD, E., A. BEYLIN, P. TANAPAT, A. REEVES, and T. J. SHORS. "Learn-ing Enhances Adult Neurogenesis in the Hippocampal Formation." *Nature Neuroscience* 2, no. 3 (1999): 260–65.

GOULD, E., A. J. REEVES, M. S. GRAZIANO, and C. G. GROSS. "Neurogenesis in the Neocortex of Adult Primates." *Science* 286, no. 5439 (1999): 548–52.

GOULD, E., P. TANAPAT, B. S. MCEWEN, G. FLUGGE, and E. FUCHS. "Proliferation of Granule Cell Precursors in the Dentate Gyrus of Adult Monkeys Is Diminished by Stress." *Proceedings of the National Academy of Sciences of the USA* 95, no. 6 (1998): 3168–71.

GOWANS, J. L., D. D. MCGREGOR, and D. M. COWEN. "Initiation of Immune Responses by Small Lymphocytes." *Nature* 196 (1962): 651–55.

GREENBERG, STEVEN. "A Concise History of Immunology." Columbia University, http://www.columbia.edu/itc/hs/medical/pathophys/immunology/readings/Concise HistoryImmunology.pdf.

GRIVENNIKOV, S. I., F. R. GRETEN, and M. KARIN. "Immunity, Inflammation, and Cancer." *Cell* 140, no. 6 (2010): 883–99.

GROHOL, JOHN. "What's the Purpose of the Fight or Flight Response?" *Psych Central* (2012), http://psychcentral.com/blog/archives/2012/12/04/whats-the-purpose-of-the-fight-or-flight-response/.

GRUVER, A. L., L. L. HUDSON, and G. D. SEMPOWSKI. "Immunosenescence of Ageing." *Journal of Pathology* 211, no. 2 (2007): 144–56.

GURIAN, MICHAEL, and KATHY STEVENS. "Educational Leadership: Closing Achievement Gaps, with Boys and Girls in Mind." Association for Supervision and Curriculum De-

velopment (ASCD) (2004), http://www.ascd.org/pub
lications/educational-leadership/nov04/vol62/num03/
With-Boys-and-Girls-in-Mind.aspx.

HALL, E. D., and J. E. SPRINGER. "Neuroprotection and Acute Spinal Cord Injury: A Reappraisal." *NeuroRx* 1, no. 1 (2004):
80–100.

HAORAH, J., B. KNIPE, J. LEIBHART, A. GHORPADE, and Y. PERSIDSKY.
"Alcohol-Induced Oxidative Stress in Brain Endothelial
Cells Causes Blood-Brain Barrier Dysfunction." *Journal of
Leukocyte Biology* 78, no. 6 (2005): 1223–32.

HARDIMAN, O., L. H. VAN DEN BERG, and M. C. KIERNAN. "Clinical
Diagnosis and Management of Amyotrophic Lateral Sclerosis." *Nature Reviews. Neurology* 7, no. 11 (2011): 639–49.

HARVARD SCHOOL OF PUBLIC HEALTH. "International Survey
Highlights Great Public Desire to Seek Early Diagnosis
of Alzheimer's" (2011), http://www.hsph.harvard.edu/
news/press-releases/alzheimers-international-survey/.

HATFIELD, GARY. "René Descartes." *The Stanford Encyclopedia of Philosophy,* http://plato.stanford.edu/archives/
sum2014/entries/descartes/.

HAUBEN, E., T. MIZRAHI, E. AGRANOV, and M. SCHWARTZ. "Sexual
Dimorphism in the Spontaneous Recovery from Spinal
Cord Injury: A Gender Gap in Beneficial Autoimmunity?"
European Journal of Neuroscience 16, no. 9 (2002): 1731–40.

"HAVING A MALE CO-TWIN IMPROVES MENTAL ROTATION PERFOR-
MANCE IN FEMALES." *ScienceDaily* (2010), www.science
daily.com/releases/2010/09/100907113046.htm.

"HENRY FORD—PRINCIPLES OF SERVICE OF A SELF-MADE MAN," http://www.henry-ford.net/english/quotes.html.

HO, A. D., W. WAGNER, and U. MAHLKNECHT. "Stem Cells and Ageing: The Potential of Stem Cells to Overcome Age-Related Deteriorations of the Body in Regenerative Medicine." *EMBO Reports* 6, special no. (2005): S35–38.

"HOW MALE AND FEMALE BRAINS DIFFER." WebMD, http://www.webmd.com/balance/features/how-male-female-brains-differ.

HSU, JEREMY. "How Much Power Does the Human Brain Require to Operate?" *Popular Science* (2009), http://www.popsci.com/technology/article/2009-11/neuron-computer-chips-could-overcome-power-limitations-digital.

HUBEL, DAVID H., and TORSTEN N. WIESEL. *Brain and Visual Perception: The Story of a 25-Year Collaboration*. New York: Oxford University Press, 2005.

HUGHES, J. T. "The Edwin Smith Surgical Papyrus: An Analysis of the First Case Reports of Spinal Cord Injuries." *Paraplegia* 26, no. 2 (1988): 71–82.

HURLBERT, R. J., M. N. HADLEY, B. C. WALTERS, B. AARABI, S. S. DHALL, D. E. GELB, C. J. ROZZELLE, T. C. RYKEN, and N. THEODORE. "Pharmacological Therapy for Acute Spinal Cord Injury." *Neurosurgery* 72, suppl. 2 (2013): 93–105.

IMMUNOBRAIN THERAPIES. http://www.immunobrain.co.il/.

ITO, Y., Y. SUGIMOTO, M. TOMIOKA, N. KAI, and M. TANAKA. "Does High Dose Methylprednisolone Sodium Succinate Really

Improve Neurological Status in Patient with Acute Cervical Cord Injury? A Prospective Study about Neurological Recovery and Early Complications." *Spine* 34, no. 20 (2009): 2121–24.

JARRETT, CHRISTIAN. "The Truth about the Effect of Pregnancy on Women's Brains." *Psychology Today* (2012), http://www.psychologytoday.com/blog/brain-myths/201208/the-truth-about-the-effect-pregnancy-womens-brains.

JIN, K., A. L. PEEL, X. O. MAO, L. XIE, B. A. COTTRELL, D. C. HENSHALL, and D. A. GREENBERG. "Increased Hippocampal Neurogenesis in Alzheimer's Disease." *Proceedings of the National Academy of Sciences of the USA* 101, no. 1 (2004): 343–47.

JUNG, S., J. ALIBERTI, P. GRAEMMEL, M. J. SUNSHINE, G. W. KREUTZBERG, A. SHER, and D. R. LITTMAN. "Analysis of Fractalkine Receptor CX(3)CR1 Function by Targeted Deletion and Green Fluorescent Protein Reporter Gene Insertion." *Molecular and Cellular Biology* 20, no. 11 (2000): 4106–14.

KANDEL, ERIC R., JAMES H. SCHWARTZ, and THOMAS M. JESSELL, EDS. *Principles of Neural Science.* New York: McGraw-Hill Medical, 2000.

KAO, C. C. "Comparison of Healing Process in Transected Spinal Cords Grafted with Autogenous Brain Tissue, Sciatic Nerve, and Nodose Ganglion." *Experimental Neurology* 44, no. 3 (1974): 424–39.

KAPLAN, M. S. "Neurogenesis in the 3-Month-Old Rat Visual Cortex." *Journal of Comparative Neurology* 195, no. 2 (1981): 323–38.

KAWAMURA, N., Y. KIM, and N. ASUKAI. "Suppression of Cellular Immunity in Men with a Past History of Posttraumatic Stress Disorder." *American Journal of Psychiatry* 158, no. 3 (2001): 484–86.

KEMPERMANN, G., H. G. KUHN, and F. H. GAGE. "More Hippocampal Neurons in Adult Mice Living in an Enriched Environment." *Nature* 386, no. 6624 (1997): 493–95.

KIECOLT-GLASER, J. K., M. A. BELURY, R. ANDRIDGE, W. B. MALARKEY, and R. GLASER. "Omega-3 Supplementation Lowers Inflammation and Anxiety in Medical Students: A Randomized Controlled Trial." *Brain, Behavior, and Immunity* 25, no. 8 (2011): 1725–34.

KIMPTON, J. "The Brain Derived Neurotrophic Factor and Influences of Stress in Depression." *Psychiatria Danubina* 24, suppl. 1 (2012): S169–71.

KIPNIS, J., and M. SCHWARTZ. "Dual Action of Glatiramer Acetate (Cop-1) in the Treatment of CNS Autoimmune and Neurodegenerative Disorders." *Trends in Molecular Medicine* 8, no. 7 (2002): 319–23.

KIPNIS, J., H. COHEN, M. CARDON, Y. ZIV, and M. SCHWARTZ. "T Cell Deficiency Leads to Cognitive Dysfunction: Implications for Therapeutic Vaccination for Schizophrenia and Other Psychiatric Conditions." *Proceedings of the National Academy of Sciences of the USA* 101, no. 21 (2004): 8180–85.

KIPNIS, J., E. YOLES, Z. PORAT, A. COHEN, F. MOR, M. SELA, I. R. COHEN, and M. SCHWARTZ. "T Cell Immunity to Copolymer 1 Confers Neuroprotection on the Damaged Optic Nerve:

Possible Therapy for Optic Neuropathies." *Proceedings of the National Academy of Sciences of the USA* 97, no. 13 (2000): 7446–51.

KNOLLER, N., G. AUERBACH, V. FULGA, G. ZELIG, J. ATTIAS, R. BAKIMER, J. B. MARDER, E. YOLES, M. BELKIN, M. SCHWARTZ, and M. HADANI. "Clinical Experience Using Incubated Autologous Macrophages as a Treatment for Complete Spinal Cord Injury: Phase I Study Results." *Journal of Neurosurgery Spine* 3, no. 3 (2005): 173–81.

KORONYO-HAMAOUI, M., M. K. KO, Y. KORONYO, D. AZOULAY, A. SEKSENYAN, G. KUNIS, M. PHAM, J. BAKHSHESHIAN, P. ROGERI, K. L. BLACK, D. L. FARKAS, and M. SCHWARTZ. "Attenuation of AD-Like Neuropathology by Harnessing Peripheral Immune Cells: Local Elevation of IL-10 and MMP-9." *Journal of Neurochemistry* 111, no. 6 (2009): 1409–24.

KRAUSE, D. L., and N. MULLER. "The Relationship between Tourette's Syndrome and Infections." *The Open Neurology Journal* 6 (2012): 124–28.

KUNIS, G., K. BARUCH, O. MILLER, and M. SCHWARTZ. "Immunization with a Myelin-Derived Antigen Activates the Brain's Choroid Plexus for Recruitment of Immunoregulatory Cells to the CNS and Attenuates Disease Progression in a Mouse Model of ALS." *Journal of Neuroscience* 35, no. 16 (2015): 6381–93.

KUNIS, G., K. BARUCH, N. ROSENZWEIG, A. KERTSER, O. MILLER, T. BERKUTZKI, and M. SCHWARTZ. "IFN-Gamma-Dependent Activation of the Brain's Choroid Plexus for CNS Immune Surveillance and Repair." *Brain* 136, pt. 11 (2013): 3427–40.

LAGES, C. S., I. SUFFIA, P. A. VELILLA, B. HUANG, G. WARSHAW, D. A. HILDEMAN, Y. BELKAID, and C. CHOUGNET. "Functional Regulatory T Cells Accumulate in Aged Hosts and Promote Chronic Infectious Disease Reactivation." *Journal of Immunology* 181, no. 3 (2008): 1835–48.

LAWLOR, CLARK. *From Melancholia to Prozac: A History of Depression*. Oxford: Oxford University Press, 2012.

LEE, B. B., R. A. CRIPPS, M. FITZHARRIS, and P. C. WING. "The Global Map for Traumatic Spinal Cord Injury Epidemiology: Update 2011, Global Incidence Rate." *Spinal Cord* 52, no. 2 (2014): 110–16.

LEE-LIU, D., G. EDWARDS-FARET, V. S. TAPIA, and J. LARRAIN. "Spinal Cord Regeneration: Lessons for Mammals from Non-Mammalian Vertebrates." *Genesis* 51, no. 8 (2013): 529–44.

LEVIN, L. A., and P. PEEPLES. "History of Neuroprotection and Rationale as a Therapy for Glaucoma." *American Journal of Managed Care* 14, no. 1, suppl. (2008): S11–14.

LEWITUS, G. M., H. COHEN, and M. SCHWARTZ. "Reducing Post-Traumatic Anxiety by Immunization." *Brain, Behavior, and Immunity* 22, no. 7 (2008): 1108–14.

LEWITUS, G. M., and M. SCHWARTZ. "Behavioral Immunization: Immunity to Self-Antigens Contributes to Psychological Stress Resilience." *Molecular Psychiatry* 14, no. 5 (2009): 532–36.

LEWITUS, G. M., A. WILF-YARKONI, Y. ZIV, M. SHABAT-SIMON, R. GERSNER, A. ZANGEN, and M. SCHWARTZ. "Vaccination as a Novel

Approach for Treating Depressive Behavior." *Biological Psychiatry* 65, no. 4 (2009): 283–88.

LIM, S., C. E. HAN, P. J. UHLHAAS, and M. KAISER. "Preferential Detachment during Human Brain Development: Age-and Sex-Specific Structural Connectivity in Diffusion Tensor Imaging (DTI) Data." *Cerebral Cortex* (2013).

LLEDO, P. M., M. ALONSO, and M. S. GRUBB. "Adult Neurogenesis and Functional Plasticity in Neuronal Circuits." *Nature Reviews. Neuroscience* 7, no. 3 (2006): 179–93.

LOKHORST, GERT-JAN. "Descartes and the Pineal Gland." *The Stanford Encyclopedia of Philosophy,* http://plato.stanford.edu/archives/spr2014/entries/pineal-gland/.

LONDON, A., I. BENHAR, and M. SCHWARTZ. "The Retina as a Window to the Brain—from Eye Research to CNS Disorders." *Nature Reviews. Neurology* 9, no. 1 (2013): 44–53.

LONDON, A., E. ITSKOVICH, I. BENHAR, V. KALCHENKO, M. MACK, S. JUNG, and M. SCHWARTZ. "Neuroprotection and Progenitor Cell Renewal in the Injured Adult Murine Retina Requires Healing Monocyte-Derived Macrophages." *Journal of Experimental Medicine* 208, no. 1 (2011): 23–39.

LOU GEHRIG. http://www.lougehrig.com/.

LUPPI, P. "How Immune Mechanisms Are Affected by Pregnancy." *Vaccine* 21, no. 24 (2003): 3352–57.

LYNCH, D. R., and T. M. DAWSON. "Secondary Mechanisms in Neuronal Trauma." *Current Opinion in Neurology* 7, no. 6 (1994): 510–16.

MACE, NANCY L., and PETER V. RABINS. *The 36-Hour Day: A Family Guide to Caring for People with Alzheimer Disease, Other Dementias, and Memory Loss in Later Life*. 4th ed. Baltimore: Johns Hopkins University Press, 2006.

MACMILLAN, AMANDA, and TAMARA SCHRYVER. "9 Power Foods That Boost Immunity." *Prevention*, n.d., http://www .prevention.com/food/healthy-eating-tips/power-foods -boost-immunity?s=1.

MALBERG, J. E., A. J. EISCH, E. J. NESTLER, and R. S. DUMAN. "Chronic Antidepressant Treatment Increases Neurogenesis in Adult Rat Hippocampus." *Journal of Neuroscience* 20, no. 24 (2000): 9104–10.

MANGELS, JENNIFER. "History of Neuroscience." Columbia University, http://www.columbia.edu/cu/psychology/ courses/1010/mangels/neuro/history/history.html.

MANTOVANI, A., P. ALLAVENA, A. SICA, and F. BALKWILL. "Cancer-Related Inflammation." *Nature* 454, no. 7203 (2008): 436–44.

MARTIN, P. "Wound Healing—Aiming for Perfect Skin Regeneration." *Science* 276, no. 5309 (1997): 75–81.

MASTERS, C. L., T. WYSS-CORAY, and G. M. PASINETTI. "Anti-Inflammatory Drugs Fall Short in Alzheimer's Disease." *Nature Medicine* 14, no. 9 (2008): 916.

MCCALL, A. L., W. R. MILLINGTON, and R. J. WURTMAN. "Blood-Brain Barrier Transport of Caffeine: Dose-Related Restriction of Adenine Transport." *Life Sciences* 31, no. 24 (1982): 2709–15.

MCCARTHY, M. M., A. P. ARNOLD, G. F. BALL, J. D. BLAUSTEIN, and G. J. DE VRIES. "Sex Differences in the Brain: The Not So Inconvenient Truth." *Journal of Neuroscience* 32, no. 7 (2012): 2241–47.

MCCOMBE, P. A., and R. D. HENDERSON. "Effects of Gender in Amyotrophic Lateral Sclerosis." *Gender Medicine* 7, no. 6 (2010): 557–70.

MCEWEN, B. S. "Physiology and Neurobiology of Stress and Adaptation: Central Role of the Brain." *Physiological Reviews* 87, no. 3 (2007): 873–904.

MCEWEN, BRUCE S., and HAROLD M. SCHMECK. *The Hostage Brain*. New York: Rockefeller University Press, 1994.

"MEDITATION IMPROVES THE IMMUNE SYSTEM, RESEARCH SHOWS." Telegraph (2011), http://www.telegraph.co.uk /health/healthnews/8862275/Meditation-improves -the-immune-system-research-shows.html.

"MELISSA'S STORY." CBSNews.com (2001), http://www .cbsnews.com/news/melissas-story-14-06-2001/.

"MEN AND WOMEN USE BRAIN DIFFERENTLY." WebMD, http:// www.webmd.com/brain/news/20060719/men -women-use-brain-differently.

MERCANDETTI, MICHAEL, ADAM J. COHEN, and JOSEPH A. MOLNAR. "Wound Healing and Repair." Medscape (2013), http:// emedicine.medscape.com/article/1298129-overview #aw2aab6b6.

MESSAOUDI, I., J. WARNER, M. FISCHER, B. PARK, B. HILL, J. MATTISON, M. A. LANE, G. S. ROTH, D. K. INGRAM, L. J. PICKER, D. C. DOUEK,

M. MORI, and J. NIKOLICH-ZUGICH. "Delay of T Cell Senescence by Caloric Restriction in Aged Long-Lived Non-
human Primates." *Proceedings of the National Academy of
Sciences of the USA* 103, no. 51 (2006): 19448–53.

MEYER, T. E., S. J. KOVACS, A. A. EHSANI, S. KLEIN, J. O. HOLLOSZY, and
L. FONTANA. "Long-Term Caloric Restriction Ameliorates
the Decline in Diastolic Function in Humans." *Journal of
the American College of Cardiology* 47, no. 2 (2006): 398–
402.

MGH CENTER FOR WOMEN'S MENTAL HEALTH. "Hormone Replacement Therapy Revisited" (2008). http://womensmental
health.org/posts/hormone-replacement-therapy-revis
ited/.

MILLER, I. N., and A. CRONIN-GOLOMB. "Gender Differences in
Parkinson's Disease: Clinical Characteristics and Cognition." *Movement Disorders* 25, no. 16 (2010): 2695–2703.

MITNE, S., S. H. TEIXEIRA, M. SCHWARTZ, M. BELKIN, M. E. FARAH,
N. S. DE MORAES, L. DA CRUZ NOIA, A. T. PAES, C. L. LOTTEN-
BERG, and A. PARANHOS JUNIOR. "The Potential Neuro-
protective Effects of Weekly Treatment with Glatiramer
Acetate in Diabetic Patients after Panretinal Photo-
coagulation." *Clinical Ophthalmology* 5 (2011): 991–97.

MOALEM, G., R. LEIBOWITZ-AMIT, E. YOLES, F. MOR, I. R. COHEN, and
M. SCHWARTZ. "Autoimmune T Cells Protect Neurons
from Secondary Degeneration after Central Nervous System Axotomy." *Nature Medicine* 5, no. 1 (1999): 49–55.

MUKHERJEE, SIDDHARTHA. "Post-Prozac Nation: The Science and

History of Treating Depression." *New York Times Magazine,* April 22, 2012, http://www.nytimes.com/2012/04/22 /magazine/the-science-and-history-of-treating-depres sion.html?pagewanted=all&_r=0.

MURPHY, J. B., and E. STURM. "Conditions Determining the Transplantability of Tissues in the Brain." *Journal of Experimental Medicine* 38, no. 2 (1923): 183–97.

NATIONAL EYE INSTITUTE, NATIONAL INSTITUTES OF HEALTH, U.S. DEPARTMENT OF HEALTH AND HUMAN SERVICES. "Facts about Glaucoma." https://www.nei.nih.gov/health/ glaucoma/glaucoma_facts.asp.

NATIONAL INSTITUTE ON AGING, NATIONAL INSTITUTES OF HEALTH, U.S. DEPARTMENT OF HEALTH AND HUMAN SERVICES. "Alzheimer's Disease: Unraveling the Mystery" (2008). http://www.nia.nih.gov/alzheimers/publication/part -3-ad-research-better-questions-new-answers/looking -causes-ad.

———. "Alzheimer's Disease Medications Fact Sheet" (2008). http://www.nia.nih.gov/alzheimers/publication/ alzheimers-disease-medications-fact-sheet.

———. "The Changing Brain in Healthy Aging." http://www .nia.nih.gov/alzheimers/publication/part-1-basics -healthy-brain/changing-brain-healthy-aging.

NATIONAL INSTITUTE OF MENTAL HEALTH, NATIONAL INSTITUTES OF HEALTH, U.S. DEPARTMENT OF HEALTH AND HUMAN SERVICES. "Brain Basics." http://www.nimh.nih.gov/health/educa tional-resources/brain-basics/brain-basics.shtml.

———. "Post-Traumatic Stress Disorder (PTSD)." http://www.nimh.nih.gov/health/publications/post-traumatic-stress-disorder-ptsd/index.shtml.

NATIONAL INSTITUTE OF NEUROLOGICAL DISORDERS AND STROKE, NATIONAL INSTITUTES OF HEALTH, U.S. DEPARTMENT OF HEALTH AND HUMAN SERVICES. "Amyotrophic Lateral Sclerosis (ALS) Fact Sheet." http://www.ninds.nih.gov/disorders/amyotrophiclateralsclerosis/detail_ALS.htm.

———. "Spinal Cord Injury: Hope through Research." http://www.ncbi.nlm.nih.gov/pubmed/.

NATIONAL INSTITUTES OF HEALTH, U.S. DEPARTMENT OF HEALTH AND HUMAN SERVICES. "Stem Cell Basics." In Stem Cell Information, http://stemcells.nih.gov/info/basics/Pages/Default.aspx.

NEMADE, RASHMI, NATALIE STAATS REISS, and MARK DOMBECK. "Depression: Major Depression and Unipolar Varieties. Historical Understandings of Depression." www.mentalhelp.net, http://www.mentalhelp.net/poc/view_doc.php?type=doc&id=12995&cn=5.

NEWBERG, A. B., N. WINTERING, D. S. KHALSA, H. ROGGENKAMP, and M. R. WALDMAN. "Meditation Effects on Cognitive Function and Cerebral Blood Flow in Subjects with Memory Loss: A Preliminary Study." *Journal of Alzheimer's Disease* 20, no. 2 (2010): 517–26.

NEWQUIST, H. P., KEITH KASNOT, and ERIC BRACE. *The Great Brain Book: An Inside Look at the Inside of Your Head.* New York: Scholastic Reference, 2004.

NIEDERKORN, J. Y. "See No Evil, Hear No Evil, Do No Evil: The Lessons of Immune Privilege." *Nature Immunology* 7, no. 4 (2006): 354–59.

NORTHOFF, GEORG. "Psychoanalysis and the Brain—Why Did Freud Abandon Neuroscience?" *Frontiers in Psychology* 3 (2012): 71.

OERTELT-PRIGIONE, S. "The Influence of Sex and Gender on the Immune Response." *Autoimmunity Reviews* 11, nos. 6–7 (2012): A479–85.

OKINAWA CENTENARIAN STUDY. http://www.okicent.org/study.html.

OLIVER, J. E., and A. J. SILMAN. "Why Are Women Predisposed to Autoimmune Rheumatic Diseases?" *Arthritis Research and Therapy* 11, no. 5 (2009): 252.

OMODEI, D., and L. FONTANA. "Calorie Restriction and Prevention of Age-Associated Chronic Disease." *FEBS Letters* 585, no. 11 (2011): 1537–42.

ORGOGOZO, J. M., S. GILMAN, J. F. DARTIGUES, B. LAURENT, M. PUEL, L. C. KIRBY, P. JOUANNY, B. DUBOIS, L. EISNER, S. FLITMAN, B. F. MICHEL, M. BOADA, A. FRANK, and C. HOCK. "Subacute Meningoencephalitis in a Subset of Patients with AD after Abeta42 Immunization." *Neurology* 61, no. 1 (2003): 46–54.

PAPPAS, STEPHANIE. "Good Stress Response Enhances Recovery from Surgery, Stanford Study Shows." Stanford University School of Medicine, http://med.stanford.edu/ism/2009/december/adapt-stress.html.

PARKINSON'S DISEASE FOUNDATION. "Understanding Parkinson's. Parkinson's FAQ." http://www.pdf.org/pdf/fs_frequently _asked_questions_10.pdf.

PERRY, V. H., P. B. ANDERSSON, and S. GORDON. "Macrophages and Inflammation in the Central Nervous System." *Trends in Neurosciences* 16, no. 7 (1993): 268–73.

PIERIBONE, VINCENT, and DAVID F. GRUBER. *Aglow in the Dark: The Revolutionary Science of Biofluorescence.* Cambridge: Belknap Press of Harvard University Press, 2005.

POPOVICH, P. G., Z. GUAN, P. WEI, I. HUITINGA, N. VAN ROOIJEN, and B. T. STOKES. "Depletion of Hematogenous Macrophages Promotes Partial Hindlimb Recovery and Neuroanatomical Repair after Experimental Spinal Cord Injury." *Experimental Neurology* 158, no. 2 (1999): 351–65.

"POSTMENOPAUSAL HORMONES: HORMONE THERAPY: WHAT HAPPENED?" Harvard Health Publications, Harvard University, http://www.health.harvard.edu/newsweek/Post menopausal_hormones_Hormone_therapy.htm.

PRONEURON BIOTECHNOLOGIES. http://www.proneuron .com/.

"PROTECTIVE AUTOIMMUNITY." Wikipedia. http://en.wikipedia .org/wiki/Protective_autoimmunity.

"PTSD: A GROWING EPIDEMIC." *NIH MedlinePlus* (2009), http:// www.nlm.nih.gov/medlineplus/magazine/issues/ winter09/articles/winter09pg10-14.html.

RADDEN, JENNIFER. *The Nature of Melancholy: From Aristotle to Kristeva.* Oxford: Oxford University Press, 2000.

RAMÓN Y CAJAL, S. *Degeneration and Regeneration of the Nervous System*. Vol. 2. New York: Haffner, 1928.

RANSOHOFF, R. M., and B. ENGELHARDT. "The Anatomical and Cellular Basis of Immune Surveillance in the Central Nervous System." *Nature Reviews. Immunology* 12, no. 9 (2012): 623–35.

RAPALINO, O., O. LAZAROV-SPIEGLER, E. AGRANOV, G. J. VELAN, E. YOLES, M. FRAIDAKIS, A. SOLOMON, R. GEPSTEIN, A. KATZ, M. BELKIN, M. HADANI, and M. SCHWARTZ. "Implantation of Stimulated Homologous Macrophages Results in Partial Recovery of Paraplegic Rats." *Nature Medicine* 4, no. 7 (1998): 814–21.

RAPOSO, C., and M. SCHWARTZ. "Glial Scar and Immune Cell Involvement in Tissue Remodeling and Repair Following Acute CNS Injuries." *Glia* 62, no. 11 (2014): 1895–1904.

RAPPOPORT, JON. "Your Mind Is Not a Computer." In Jon Rappoport's Blog (2013), https://jonrappoport.wordpress.com/2013/06/03/your-mind-is-not-a-computer/.

RATEY, JOHN J., and ERIC HAGERMAN. *Spark: The Revolutionary New Science of Exercise and the Brain*. New York: Little, Brown, 2008.

RENTON, A. E., A. CHIO, and B. J. TRAYNOR. "State of Play in Amyotrophic Lateral Sclerosis Genetics." *Nature Neuroscience* 17, no. 1 (2014): 17–23.

RENTON, A. E., E. MAJOUNIE, A. WAITE, J. SIMON-SANCHEZ, S. ROLLINSON, J. R. GIBBS, J. C. SCHYMICK, H. LAAKSOVIRTA, J. C. VAN SWIETEN, L. MYLLYKANGAS, H. KALIMO, A. PAETAU, Y. ABRAMZON,

A. M. REMES, A. KAGANOVICH, S. W. SCHOLZ, J. DUCKWORTH, J. DING, D. W. HARMER, D. G. HERNANDEZ, J. O. JOHNSON, K. MOK, M. RYTEN, D. TRABZUNI, R. J. GUERREIRO, R. W. ORRELL, J. NEAL, A. MURRAY, J. PEARSON, I. E. JANSEN, D. SONDERVAN, H. SEELAAR, D. BLAKE, K. YOUNG, N. HALLIWELL, J. B. CALLISTER, G. TOULSON, A. RICHARDSON, A. GERHARD, J. SNOWDEN, D. MANN, D. NEARY, M. A. NALLS, T. PEURALINNA, L. JANSSON, V. M. ISOVIITA, A. L. KAIVORINNE, M. HOLTTA-VUORI, E. IKONEN, R. SULKAVA, M. BENATAR, J. WUU, A. CHIO, G. RESTAGNO, G. BORGHERO, M. SABATELLI, ITALSGEN CONSORTIUM, D. HECK-ERMAN, E. ROGAEVA, L. ZINMAN, J. D. ROTHSTEIN, M. SENDTNER, C. DREPPER, E. E. EICHLER, C. ALKAN, Z. ABDULLAEV, S. D. PACK, A. DUTRA, E. PAK, J. HARDY, A. SINGLETON, N. M. WILLIAMS, P. HEUTINK, S. PICKERING-BROWN, H. R. MORRIS, P. J. TIENARI, and B. J. TRAYNOR. "A Hexanucleotide Repeat Expansion in C9ORF72 Is the Cause of Chromosome 9p21-Linked ALS-FTD." *Neuron* 72, no. 2 (2011): 257–68.

RICHARDSON, P. M., U. M. MCGUINNESS, and A. J. AGUAYO. "Axons from CNS Neurons Regenerate into PNS Grafts." *Nature* 284, no. 5753 (1980): 264–65.

RIPPS, M. E., G. W. HUNTLEY, P. R. HOF, J. H. MORRISON, and J. W. GOR-DON. "Transgenic Mice Expressing an Altered Murine Superoxide Dismutase Gene Provide an Animal Model of Amyotrophic Lateral Sclerosis." *Proceedings of the National Academy of Sciences of the USA* 92, no. 3 (1995): 689–93.

ROCHELLE, OLIVER. "Traumatic Experiences Weaken Immune-System Gene." *Psych Central* (2010), http://psychcentral

.com/news/2010/05/04/traumatic-experiences-weaken
-immune-system-gene/13497.html.

ROGERS, C. J., D. A. ZAHAROFF, K. W. HANCE, S. N. PERKINS, S. D. HURS-
TING, J. SCHLOM, and J. W. GREINER. "Exercise Enhances Vaccine-Induced Antigen-Specific T Cell Responses." *Vaccine*
26, no. 42 (2008): 5407–15.

ROLLS, A., H. SCHORI, A. LONDON, and M. SCHWARTZ. "Decrease in
Hippocampal Neurogenesis During Pregnancy: A Link to
Immunity." *Molecular Psychiatry* 13, no. 5 (2008): 468–69.

ROLLS, A., R. SHECHTER, A. LONDON, Y. SEGEV, J. JACOB-HIRSCH,
N. AMARIGLIO, G. RECHAVI, and M. SCHWARTZ. "Two Faces of
Chondroitin Sulfate Proteoglycan in Spinal Cord Repair:
A Role in Microglia/Macrophage Activation." *PLoS Medi-
cine* 5, no. 8 (2008): e171.

ROLLS, A., R. SHECHTER, and M. SCHWARTZ. "The Bright Side of
the Glial Scar in CNS Repair." *Nature Reviews. Neuroscience*
10, no. 3 (2009): 235–41.

RON-HAREL, N., and M. SCHWARTZ. "Immune Senescence and
Brain Aging: Can Rejuvenation of Immunity Reverse
Memory Loss?" *Trends in Neurosciences* 32, no. 7 (2009):
367–75.

RON-HAREL, N., Y. SEGEV, G. M. LEWITUS, M. CARDON, Y. ZIV, D. NETA-
NELY, J. JACOB-HIRSCH, N. AMARIGLIO, G. RECHAVI, E. DOMANY,
and M. SCHWARTZ. "Age-Dependent Spatial Memory Loss
Can Be Partially Restored by Immune Activation." *Rejuvenation Research* 11, no. 5 (2008): 903–13.

ROSEN, D. R., T. SIDDIQUE, D. PATTERSON, D. A. FIGLEWICZ, P. SAPP,

A. HENTATI, D. DONALDSON, J. GOTO, J. P. O'REGAN, H. X. DENG, and ET AL. "Mutations in Cu/Zn Superoxide Dismutase Gene Are Associated with Familial Amyotrophic Lateral Sclerosis." *Nature* 362, no. 6415 (1993): 59–62.

ROSENBERG, S. A., J. C. YANG, D. J. SCHWARTZENTRUBER, P. HWU, F. M. MARINCOLA, S. L. TOPALIAN, N. P. RESTIFO, M. E. DUDLEY, S. L. SCHWARZ, P. J. SPIESS, J. R. WUNDERLICH, M. R. PARKHURST, Y. KAWAKAMI, C. A. SEIPP, J. H. EINHORN, and D. E. WHITE. "Immunologic and Therapeutic Evaluation of a Synthetic Peptide Vaccine for the Treatment of Patients with Metastatic Melanoma." *Nature Medicine* 4, no. 3 (1998): 321–27.

ROTHSTEIN, EDWARD. "'Dialog in the Dark' at South Street Seaport—Exhibition Review." *New York Times*, August 19, 2011, http://www.nytimes.com/2011/08/19/arts/design/dialog-in-the-dark-at-south-street-seaport-exhibition-review.html.

SALTSMAN, KIRSTIE. "Chapter 5: The Last Chapter; Cell Aging and Death; Inside the Cell." National Institutes of Health, U.S. Department of Health and Human Services, http://publications.nigms.nih.gov/insidethecell/chapter5.html.

SAPOLSKY, ROBERT M. *Why Zebras Don't Get Ulcers*. 3rd ed. New York: Times Books, 2004.

SAUL, A. N., T. M. OBERYSZYN, C. DAUGHERTY, D. KUSEWITT, S. JONES, S. JEWELL, W. B. MALARKEY, A. LEHMAN, S. LEMESHOW, and F. S. DHABHAR. "Chronic Stress and Susceptibility to Skin

Cancer." *Journal of the National Cancer Institute* 97, no. 23 (2005): 1760–67.

SCHATZ, PHILIP. "History of Neuropsychology." Saint Joseph's University, http://schatz.sju.edu/neuro/nphistory/nphistory.html.

SCHILLER, M. D., and R. J. MOBBS. "The Historical Evolution of the Management of Spinal Cord Injury." *Journal of Clinical Neuroscience* 19, no. 10 (2012): 1348–53.

SCHORI, H., E. YOLES, and M. SCHWARTZ. "T-Cell-Based Immunity Counteracts the Potential Toxicity of Glutamate in the Central Nervous System." *Journal of Neuroimmunology* 119, no. 2 (2001): 199–204.

SCHWAB, M. E. "Functions of Nogo Proteins and Their Receptors in the Nervous System." *Nature Reviews. Neuroscience* 11, no. 12 (2010): 799–811.

SCHWARTZ, M. "Harnessing the Immune System for Neuroprotection: Therapeutic Vaccines for Acute and Chronic Neurodegenerative Disorders." *Cellular and Molecular Neurobiology* 21, no. 6 (2001): 617–27.

———. "Lecture by Professor Michal Schwartz: 'The Immune System Is Needed for Shaping, Protecting, and Healing the Brain.'" Australian National University channel, YouTube, https://www.youtube.com/watch?v=7W j7EX_mb20.

———. "Neurodegeneration and Neuroprotection in Glaucoma: Development of a Therapeutic Neuroprotective

Vaccine: The Friedenwald Lecture." *Investigative Ophthalmology and Visual Science* 44, no. 4 (2003): 1407–11.

———. "Physiological Approaches to Neuroprotection: Boosting of Protective Autoimmunity." *Survey of Ophthalmology* 45, suppl. 3 (2001): S256–60; discussion S273–76.

———. "Vaccination for Glaucoma: Dream or Reality?" *Brain Research Bulletin* 62, no. 6 (2004): 481–84.

SCHWARTZ, M., and K. BARUCH. "The Resolution of Neuroinflammation in Neurodegeneration: Leukocyte Recruitment via the Choroid Plexus." *EMBO Journal* 33, no. 1 (2014): 7–22.

———. "Vaccine for the Mind: Immunity against Self at the Choroid Plexus for Erasing Biochemical Consequences of Stressful Episodes." *Human Vaccines and Immunotherapeutics* 8, no. 10 (2012): 1465–68.

SCHWARTZ, M., M. BELKIN, A. HAREL, A. SOLOMON, V. LAVIE, M. HADANI, I. RACHAILOVICH, and C. STEIN-IZSAK. "Regenerating Fish Optic Nerves and a Regeneration-Like Response in Injured Optic Nerves of Adult Rabbits." *Science* 228, no. 4699 (1985): 600–603.

SCHWARTZ, M., M. BELKIN, E. YOLES, and A. SOLOMON. "Potential Treatment Modalities for Glaucomatous Neuropathy: Neuroprotection and Neuroregeneration." *Journal of Glaucoma* 5, no. 6 (1996): 427–32.

SCHWARTZ, M., and J. KIPNIS. "Model of Acute Injury to Study

Neuroprotection." *Methods in Molecular Biology* 399 (2007): 41–53.

———. "Protective Autoimmunity and Neuroprotection in Inflammatory and Noninflammatory Neurodegenerative Diseases." *Journal of the Neurological Sciences* 233, nos. 1–2 (2005): 163–66.

SCHWARTZ, M., and C. RAPOSO. "Protective Autoimmunity: A Unifying Model for the Immune Network Involved in CNS Repair." *Neuroscientist* 20, no. 4 (2014): 343–58.

SCHWARTZ, M., I. SHAKED, J. FISHER, T. MIZRAHI, and H. SCHORI. "Protective Autoimmunity against the Enemy Within: Fighting Glutamate Toxicity." *Trends in Neurosciences* 26, no. 6 (2003): 297–302.

SCHWARTZ, M., and R. SHECHTER. "Protective Autoimmunity Functions by Intracranial Immunosurveillance to Support the Mind: The Missing Link between Health and Disease." *Molecular Psychiatry* 15, no. 4 (2010): 342–54.

———. "Systemic Inflammatory Cells Fight Off Neurodegenerative Disease." *Nature Reviews. Neurology* 6, no. 7 (2010): 405–10.

SCHWARTZ, M., and E. YOLES. "Neuroprotection: A New Treatment Modality for Glaucoma?" *Current Opinion in Ophthalmology* 11, no. 2 (2000): 107–11.

———. "Optic Nerve Degeneration and Potential Neuroprotection: Implications for Glaucoma." *European Journal of Ophthalmology* 9, suppl. 1 (1999): S9–11.

SCHWARTZ, M., and Y. ZIV. "Immunity to Self and Self-Maintenance: A Unified Theory of Brain Pathologies." *Trends in Immunology* 29, no. 5 (2008): 211–19.

———. "Immunity to Self and Self-Maintenance: What Can Tumor Immunology Teach Us about ALS and Alzheimer's Disease?" *Trends in Pharmacological Sciences* 29, no. 6 (2008): 287–93.

SCHWARZ, JACLYN M., and STACI D. BILBO. *The Immune System and the Developing Brain.* San Rafael, Calif.: Morgan and Claypool, 2011.

SEGMAN, R. H., N. SHEFI, T. GOLTSER-DUBNER, N. FRIEDMAN, N. KAMINSKI, and A. Y. SHALEV. "Peripheral Blood Mononuclear Cell Gene Expression Profiles Identify Emergent Post-Traumatic Stress Disorder among Trauma Survivors." *Molecular Psychiatry* 10, no. 5 (2005): 500–513.

SEKSENYAN, A., N. RON-HAREL, D. AZOULAY, L. CAHALON, M. CARDON, P. ROGERI, M. K. KO, M. WEIL, S. BULVIK, G. RECHAVI, N. AMARIGLIO, E. KONEN, M. KORONYO-HAMAOUI, R. SOMECH, and M. SCHWARTZ. "Thymic Involution, a Co-Morbidity Factor in Amyotrophic Lateral Sclerosis." *Journal of Cellular and Molecular Medicine* 14, no. 10 (2010): 2470–82.

SEROT, J. M., M. C. BENE, and G. C. FAURE. "Choroid Plexus, Aging of the Brain, and Alzheimer's Disease." *Frontiers in Bioscience* 8 (2003): s515–21.

SHECHTER, R., A. LONDON, and M. SCHWARTZ. "Orchestrated Leukocyte Recruitment to Immune-Privileged Sites: Absolute Barriers versus Educational Gates." *Nature Reviews. Immunology* 13, no. 3 (2013): 206–18.

SHECHTER, R., A. LONDON, C. VAROL, C. RAPOSO, M. CUSIMANO, G. YOVEL, A. ROLLS, M. MACK, S. PLUCHINO, G. MARTINO, S. JUNG, and M. SCHWARTZ. "Infiltrating Blood-Derived Macrophages Are Vital Cells Playing an Anti-Inflammatory Role in Recovery from Spinal Cord Injury in Mice." *PLoS Medicine* 6, no. 7 (2009): e1000113.

SHECHTER, R., O. MILLER, G. YOVEL, N. ROSENZWEIG, A. LONDON, J. RUCKH, K. W. KIM, E. KLEIN, V. KALCHENKO, P. BENDEL, S. A. LIRA, S. JUNG, and M. SCHWARTZ. "Recruitment of Beneficial M2 Macrophages to Injured Spinal Cord Is Orchestrated by Remote Brain Choroid Plexus." *Immunity* 38, no. 3 (2013): 555–69.

SHECHTER, R., C. RAPOSO, A. LONDON, I. SAGI, and M. SCHWARTZ. "The Glial Scar–Monocyte Interplay: A Pivotal Resolution Phase in Spinal Cord Repair." *PLoS One* 6, no. 12 (2011): e27969.

SHECHTER, R., and M. SCHWARTZ. "Harnessing Monocyte-Derived Macrophages to Control Central Nervous System Pathologies: No Longer 'If' but 'How.'" *Journal of Pathology* 229, no. 2 (2013): 332–46.

SHIRAI, Y. "On the Transplantation of the Rat Sarcoma in Adult Heterogenous Animals." *Japan Medical World* (1921): 14–15.

SHORT, D. J., W. S. EL MASRY, and P. W. JONES. "High Dose Methylprednisolone in the Management of Acute Spinal Cord

Injury—a Systematic Review from a Clinical Perspective." *Spinal Cord* 38, no. 5 (2000): 273–86.

SILBERMAN, D. M., V. AYELLI-EDGAR, M. ZORRILLA-ZUBILETE, L. M. ZIEHER, and A. M. GENARO. "Impaired T-Cell Dependent Humoral Response and Its Relationship with T Lymphocyte Sensitivity to Stress Hormones in a Chronic Mild Stress Model of Depression." *Brain, Behavior, and Immunity* 18, no. 1 (2004): 81–90.

SILVER, J., and J. H. MILLER. "Regeneration beyond the Glial Scar." *Nature Reviews. Neuroscience* 5, no. 2 (2004): 146–56.

SILVERSTEIN, A. M. "Ilya Metchnikoff, the Phagocytic Theory, and How Things Often Work in Science." *Journal of Leukocyte Biology* 90, no. 3 (2011): 409–10.

SILVERSTEIN, MARILYN. "Christopher Reeve: Looking Back on Israel, Looking Forward to a Cure." *New Jersey Jewish News* (2003), http://www.njjewishnews.com/njjn.com/12403/njchristopher.html.

SIMARD, A. R., D. SOULET, G. GOWING, J. P. JULIEN, and S. RIVEST. "Bone Marrow-Derived Microglia Play a Critical Role in Restricting Senile Plaque Formation in Alzheimer's Disease." *Neuron* 49, no. 4 (2006): 489–502.

"SIR FRANK MACFARLANE BURNET, 1899–1985." *Nature Immunology* 8, no. 10 (2007): 1009.

"SLIDESHOW: 15 IMMUNE BOOSTING FOODS." WebMD, http://www.webmd.com/cold-and-flu/ss/slideshow-immune-foods.

SMITH, A. K., K. N. CONNEELY, V. KILARU, K. B. MERCER, T. E. WEISS,

B. BRADLEY, Y. TANG, C. F. GILLESPIE, J. F. CUBELLS, and K. J. RESS-LER. "Differential Immune System DNA Methylation and Cytokine Regulation in Post-Traumatic Stress Disorder." *American Journal of Medical Genetics. Part B, Neuropsychiatric Genetics* 156B, no. 6 (2011): 700–708.

SMITH, M. A., S. MAKINO, R. KVETNANSKY, and R. M. POST. "Stress and Glucocorticoids Affect the Expression of Brain-Derived Neurotrophic Factor and Neurotrophin-3 mRNAs in the Hippocampus." *Journal of Neuroscience* 15, no. 3, pt. 1 (1995): 1768–77.

SOCIETY FOR NEUROSCIENCE. *Brain Facts: A Primer on the Brain and Nervous System.* 7th ed. Washington, D.C.: Society for Neuroscience, 2012.

SOMPAYRAC, LAUREN. *How the Immune System Works.* 4th ed. Chichester, West Sussex: Wiley-Blackwell, 2012.

SPIVAK, B., B. SHOHAT, R. MESTER, S. AVRAHAM, I. GIL-AD, A. BLEICH, A. VALEVSKI, and A. WEIZMAN. "Elevated Levels of Serum Interleukin-1 Beta in Combat-Related Posttraumatic Stress Disorder." *Biological Psychiatry* 42, no. 5 (1997): 345–48.

STEIN, D. G. "Brain Damage, Sex Hormones, and Recovery: A New Role for Progesterone and Estrogen?" *Trends in Neurosciences* 24, no. 7 (2001): 386–91.

STEVENS, R. D., A. BHARDWAJ, J. R. KIRSCH, and M. A. MIRSKI. "Critical Care and Perioperative Management in Traumatic Spinal Cord Injury." *Journal of Neurosurgical Anesthesiology* 15, no. 3 (2003): 215–29.

STILES, J., and T. L. JERNIGAN. "The Basics of Brain Development." *Neuropsychology Review* 20, no. 4 (2010): 327–48.

STREILEIN, J. W. "Ocular Immune Privilege: Therapeutic Opportunities from an Experiment of Nature." *Nature Reviews. Immunology* 3, no. 11 (2003): 879–89.

STROMINGER, NORMAN L., ROBERT J. DEMAREST, and LOIS LAEMLE. *Noback's Human Nervous System: Structure and Function.* 7th ed. Totowa, N.J.: Humana, 2012.

STYRON, WILLIAM. *Darkness Visible: A Memoir of Madness.* New York: Random House, 1990.

SUMMERS, LAWRENCE H. "Lawrence (Larry) Summers on Women in Science." Women in Science and Engineering Leadership Institute (WISELI), http://wiseli.engr.wisc.edu/archives/summers.php#conference-info.

———. "Remarks at NBER Conference on Diversifying the Science and Engineering Workforce." Harvard University, Office of the President, http://www.harvard.edu/president/speeches/summers_2005/nber.php.

TARLACH, GEMMA. "Why Women's Immune Systems Are Stronger than Men's." *Discover,* December 23, 2013, http://blogs.discovermagazine.com/d-brief/2013/12/23/why-womens-immune-systems-are-more-sensitive-than-mens/.

TAUBER, A. I. "Metchnikoff and the Phagocytosis Theory." *Nature Reviews. Molecular Cell Biology* 4, no. 11 (2003): 897–901.

THIES, W., L. BLEILER, and ALZHEIMER'S ASSOCIATION. "2013 Alzheimer's Disease Facts and Figures." *Alzheimer's and Dementia* 9, no. 2 (2013): 208–45.

THOMPSON, DENNIS. "'Brain Training' Benefits Seen 10 Years Later in Elderly." *Health.com*, January 14, 2014, http://news.health.com/2014/01/14/gains-of-brain-training-for-elderly-seen-10-years-later/.

TOFTGÅRD, RUNE. "Maintenance of Chromosomes by Telomeres and the Enzyme Telomerase." Nobel Prize in Physiology or Medicine, 2009, http://nobelprize.org (2009).

TREDE, K. "150 Years of Freud-Kraepelin Dualism." *Psychiatric Quarterly* 78, no. 3 (2007): 237–40.

UDDIN, M., A. E. AIELLO, D. E. WILDMAN, K. C. KOENEN, G. PAWELEC, R. DE LOS SANTOS, E. GOLDMANN, and S. GALEA. "Epigenetic and Immune Function Profiles Associated with Posttraumatic Stress Disorder." *Proceedings of the National Academy of Sciences of the USA* 107, no. 20 (2010): 9470–75.

UNIVERSITY OF MARYLAND MEDICAL SYSTEM. "Omega-3 Fatty Acids." http://umm.edu/health/medical/altmed/supplement/omega3-fatty-acids.

VAKNIN, I., G. KUNIS, O. MILLER, O. BUTOVSKY, S. BUKSHPAN, D. R. BEERS, J. S. HENKEL, E. YOLES, S. H. APPEL, and M. SCHWARTZ. "Excess Circulating Alternatively Activated Myeloid (M2) Cells Accelerate ALS Progression while Inhibiting Experimental Autoimmune Encephalomyelitis." *PLoS One* 6, no. 11 (2011): e26921.

VALIAN, VIRGINIA. "Raise Your Hand if You're a Woman in Science." *Washington Post*, January 29, 2005, http://www.washingtonpost.com/wp-dyn/articles/A46421-2005Jan29.html.

"VA SECRETARY ESTABLISHES ALS AS A PRESUMPTIVE COMPENSABLE

ILLNESS." News Releases—Office of Public and Intergovernmental Affairs, http://www.va.gov/opa/pressrel/pressrelease.cfm?id=1583.

VELASQUEZ-MANOFF, MOISES. "An Immune Disorder at the Root of Autism." *New York Times,* August 26, 2012, http://www.nytimes.com/2012/08/26/opinion/sunday/immune-disorders-and-autism.html?pagewanted=all.

VERGHESE, J., R. B. LIPTON, M. J. KATZ, C. B. HALL, C. A. DERBY, G. KUSLANSKY, A. F. AMBROSE, M. SLIWINSKI, and H. BUSCHKE. "Leisure Activities and the Risk of Dementia in the Elderly." *New England Journal of Medicine* 348, no. 25 (2003): 2508–16.

VIKHANSKI, LUBA. *In Search of the Lost Cord: Solving the Mystery of Spinal Cord Regeneration.* Washington, D.C.: Joseph Henry, 2001.

VILLEDA, S. A., J. LUO, K. I. MOSHER, B. ZOU, M. BRITSCHGI, G. BIERI, T. M. STAN, N. FAINBERG, Z. DING, A. EGGEL, K. M. LUCIN, E. CZIRR, J. S. PARK, S. COUILLARD-DESPRES, L. AIGNER, G. LI, E. R. PESKIND, J. A. KAYE, J. F. QUINN, D. R. GALASKO, X. S. XIE, T. A. RANDO, and T. WYSS-CORAY. "The Ageing Systemic Milieu Negatively Regulates Neurogenesis and Cognitive Function." *Nature* 477, no. 7362 (2011): 90–94.

"VISUAL IMPAIRMENT AND BLINDNESS." WHO Media Centre, http://www.who.int/mediacentre/factsheets/fs282/en/.

VOSS, GRETCHEN. "The Risks of Anti-Aging Medicine." *Health .com* (2011), http://www.health.com/health/article/0,,20544045,00.html.

WAGSTAFF, ANNA. "Therapeutic Cancer Vaccines: There's a New Kid on the Block." *Cancerworld*, no. 39 (2010), http://www.cancerworld.org/Articles/Issues/39/November-December-2010/Cutting-Edge/441/Therapeutic-cancer-vaccines--theres-a-new-kid-on-the-block.html.

WEINREB, R. N., and L. A. LEVIN. "Is Neuroprotection a Viable Therapy for Glaucoma?" *Archives of Ophthalmology* 117, no. 11 (1999): 1540–44.

WEINTRAUB, ARLENE. *Selling the Fountain of Youth: How the Anti-Aging Industry Made a Disease out of Getting Old—and Made Billions*. New York: Basic, 2010.

WEISSKOPF, M. G., E. J. O'REILLY, M. L. MCCULLOUGH, E. E. CALLE, M. J. THUN, M. CUDKOWICZ, and A. ASCHERIO. "Prospective Study of Military Service and Mortality from ALS." *Neurology* 64, no. 1 (2005): 32–37.

WEISSMANN, G. "The Experimental Pathology of Stress: Hans Selye to Paris Hilton." *FASEB Journal* 21, no. 11 (2007): 2635–38.

WEIZMANN INSTITUTE OF SCIENCE. "Christopher Reeve Comes to Weizmann Institute." http://wis-wander.weizmann.ac.il/christopher-reeve-comes-to-weizmann#.U3s4e5hBRdg.

WERNER, E. B., and S. M. DRANCE. "Progression of Glaucomatous Field Defects despite Successful Filtration." *Canadian Journal of Ophthalmology* 12, no. 4 (1977): 275–80.

"WHAT ARE SOME RISKS OF STEM CELL THERAPIES?" Nature Reports Stem Cells (2007).

WHITACRE, C. C. "Sex Differences in Autoimmune Disease." *Nature Immunology* 2, no. 9 (2001): 777–80.

WHITE, MICHAEL. "Girls' Immune Systems Rule, Boys Drool." *Pacific Standard* (2014), http://www.psmag.com/naviga tion/health-and-behavior/girls-immune-systems-rule -boys-drool-73250/.

WILLCOX, B. J., D. C. WILLCOX, H. TODORIKI, A. FUJIYOSHI, K. YANO, Q. HE, J. D. CURB, and M. SUZUKI. "Caloric Restriction, the Traditional Okinawan Diet, and Healthy Aging: The Diet of the World's Longest-Lived People and Its Potential Impact on Morbidity and Life Span." *Annals of the New York Academy of Sciences* 1114 (2007): 434–55.

WINBLAD, B., N. ANDREASEN, L. MINTHON, A. FLOESSER, G. IMBERT, T. DUMORTIER, R. P. MAGUIRE, K. BLENNOW, J. LUNDMARK, M. STAUFENBIEL, J. M. ORGOGOZO, and A. GRAF. "Safety, Tolerability, and Antibody Response of Active Abeta Immunotherapy with Cad106 in Patients with Alzheimer's Disease: Randomised, Double-Blind, Placebo-Controlled, First-in-Human Study." *Lancet. Neurology* 11, no. 7 (2012): 597–604.

WURTZ, R. H. "Retrospective. David H. Hubel (1926–2013)." *Science* 342, no. 6158 (2013): 572.

YOLES, E., E. HAUBEN, O. PALGI, E. AGRANOV, A. GOTHILF, A. COHEN, V. KUCHROO, I. R. COHEN, H. WEINER, and M. SCHWARTZ. "Protective Autoimmunity Is a Physiological Response to CNS Trauma." *Journal of Neuroscience* 21, no. 11 (2001): 3740–48.

YOLES, E., and M. SCHWARTZ. "Degeneration of Spared Axons

Following Partial White Matter Lesion: Implications for Optic Nerve Neuropathies." *Experimental Neurology* 153, no. 1 (1998): 1–7.

———. "Potential Neuroprotective Therapy for Glaucomatous Optic Neuropathy." *Survey of Ophthalmology* 42, no. 4 (1998): 367–72.

YOUNG, W. "Fear of Hope." *Science* 277, no. 5334 (1997): 1907.

YOUNG, W., and E. S. FLAMM. "Effect of High-Dose Corticosteroid Therapy on Blood Flow, Evoked Potentials, and Extracellular Calcium in Experimental Spinal Injury." *Journal of Neurosurgery* 57, no. 5 (1982): 667–73.

ZEA, A. H., P. C. RODRIGUEZ, M. B. ATKINS, C. HERNANDEZ, S. SIGNORETTI, J. ZABALETA, D. MCDERMOTT, D. QUICENO, A. YOUMANS, A. O'NEILL, J. MIER, and A. C. OCHOA. "Arginase-Producing Myeloid Suppressor Cells in Renal Cell Carcinoma Patients: A Mechanism of Tumor Evasion." *Cancer Research* 65, no. 8 (2005): 3044–48.

ZEIDAN, F., S. K. JOHNSON, B. J. DIAMOND, Z. DAVID, and P. GOOLKASIAN. "Mindfulness Meditation Improves Cognition: Evidence of Brief Mental Training." *Consciousness and Cognition* 19, no. 2 (2010): 597–605.

ZIMMER, CARL. "Gene Therapy Emerges from Disgrace to Be the Next Big Thing, Again." *WIRED*, August 2013, http://www.wired.com/2013/08/the-fall-and-rise-of-gene-therapy-2/.

ZIPP, F., and O. AKTAS. "The Brain as a Target of Inflammation: Common Pathways Link Inflammatory and Neurodegenerative Diseases." *Trends in Neurosciences* 29, no. 9 (2006): 518–27.

ZIV, Y., N. RON, O. BUTOVSKY, G. LANDA, E. SUDAI, N. GREENBERG, H. COHEN, J. KIPNIS, and M. SCHWARTZ. "Immune Cells Contribute to the Maintenance of Neurogenesis and Spatial Learning Abilities in Adulthood." *Nature Neuroscience* 9, no. 2 (2006): 268–75.

INDEX

Page numbers in *italics* refer to illustrations.